Growth Hormone Treatment in Down's Syndrome

Growth Hormone Treatment in Down's Syndrome

Edited by

S. CASTELLS

Department of Pediatrics, State University of New York Health Science Center at Brooklyn, New York, USA

and

K. E. WISNIEWSKI

New York State Institute for Basic Research in Developmental Disabilities, New York, USA

JOHN WILEY & SONS

Chichester · New York · Brisbane · Toronto · Singapore

Other Wiley Editorial Offices

John Wiley & Sons, Inc., 605 Third Avenue, New York, NY 10158-0012, USA

Jacaranda Wiley Ltd, G.P.O. Box 859, Brisbane, Queensland 4001, Australia

John Wiley & Sons (Canada) Ltd, 22 Worcester Road, Rexdale, Ontario M9W 1L1, Canada.

John Wiley & Sons (SEA) Pte Ltd, 37 Jalan Pemimpin #05-04, Block B, Union Industrial Building, Singapore 2057

Library of Congress Cataloging-in-Publication Data

Growth hormone treatment in Down's syndrome / edited by S. Castells and K.E. Wisniewski.
p. cm.
Includes bibliographical references and index.
ISBN 0 471 93991 9
1. Down's syndrome—Hormone therapy—Congresses. 2. Somatotropin—Therapeutic use—Congresses. 3. Down's syndrome—Endocrine aspects—Congresses. I. Castells, Salvador. II. Wisniewski, K. E. (Krystyna E.)
[DNLM: 1. Down Syndrome—drug therapy—congresses. 2. Growth Disorders—drug therapy—congresses. 3. Somatotropin—therapeutic use—congresses. WS 107 G884 1993]
RJ506.D68G76 1993
616.85′8842061—dc20
DNLM/DCL
for Library of Congress 93-8310
CIP

British Library Cataloguing in Publication Data

A catalogue record for this book is available from the British Library

ISBN 0 471 93991 9

Typeset in 10/12pt Times by Mathematical Composition Setters Ltd., Salisbury
Printed and bound in Great Britain by Biddles Ltd, Guildford and King's Lynn

Contents

Contributors

A. Alberti
Department of Pediatrics, OASI Institute (IRCCS), Via Conte Ruggero 73, 94018 Troina (EN), Italy

G. Albertini
Ospedale Pediatrico 'Bambino Gesù', Sezione Palladoro e S. Marinella AB, Via P. Foelngo 74, Rome 00165, Italy

G. Annerén
Department of Clinical Genetics, University Hospital, S-751 85 Uppsala, Sweden

I. Beaulieu
New York State Institute for Basic Research in Developmental Disabilities, 1050 Forest Hill Road, Staten Island, NY 10314, USA

M. Bobinski
New York State Institute for Basic Research in Developmental Disabilities, 1050 Forest Hill Road, Staten Island, NY 10314, USA

P. Borrelli
Reparto di Endocrinologia, Ospedale Pediatrico 'Bambino Gesù', IRCCS, Piazza S. Onofrio 4, 00165 Rome, Italy

W.T. Brown
New York State Institute for Basic Research in Developmental Disabilities, 1050 Forest Hill Road, Staten Island, NY 10314, USA

C. Carlsson-Skwirut
Department of Pathology, Karolinska Institute, Stockholm S-104 01, Sweden

S. Castells
State University of New York Health Science Center at Brooklyn, Department of Pediatrics, Box 49, 450 Clarkson Avenue, Brooklyn, NY 11203, USA

P. Ciampalini
Endocrinologaa I, Università 'La Sapienza', Via del Policlinico, 00161 Rome, Italy

F. Colabucci
Department of Pediatrics, Universita Cattolica Del Sacro Cuore, Largo Agostino Gemelli 8, 00168 Rome, Italy

A. Crinò
Reparto di Endocrinologia, Ospedale Pediatrico 'Bambino Gesù', IRCCS, Piazza S. Onofrio 4, 00165 Rome, Italy

C.E. Cronk
Department of Anthropology, Southern Illinois University, Carbondale, IL 62901, USA

M.C. DiGilio
Reparto di Endocrinologia, Servizio di Genetica Medica, Ospedale Pediatrico 'Bambino Gesù', IRCCS, Piazza S. Onofrio 4, 00165 Rome, Italy

D.L. Duffey
Central Florida Down's Syndrome Association, Inc., 3015 Lakeshore Drive, Mount Dora, FL 32757, USA

N. Fabris
INRCA, Department of Geriatrics and Gerontological Research, Via Birarelli 8, 60100, Ancona, Italy

J. Flórez
Department of Pediatrics, Faculty of Medicine, University of Cantabria, Avda. Herrera Oria s/n, 39011 Santander, Spain

L.V. García
Department of Pediatrics, Faculty of Medicine, University of Cantabria, Avda. Herrera Oria s/n, 39011 Santander, Spain

A. Giannotti
Reparto di Endocrinologia, Servizio di Genetica Medica, Ospedale Pediatrico 'Bambino Gesù', IRCCS, Piazza S. Onofrio 4, 00165 Rome, Italy

J. Gustafsson
Department of Pediatrics, University Hospital, S-751 85 Uppsala, Sweden

A. Hurwitz
City University of New York, City College, School of Nursing, Harris 312, Convent Avenue, NY 10031, USA

B. Klinger
Institute of Pediatric and Adolescent Endocrinology, Children's Medical Center of Israel, Beilinson Campus, 49100 Petah Tikva, Israel

Z. Laron
Institute of Pediatric and Adolescent Endocrinology, Children's Medical Center of Israel, Beilinson Campus, 49100 Petah Tikva, Israel

L. Lazar
Institute of Pediatric and Adolescent Endocrinology, Children's Medical Center of Israel, Beilinson Campus, 49100 Petah Tikva, Israel

M.H. Lee
New York State Institute for Basic Research in Developmental Disabilities, Department of Pathological Neurobiology, 1050 Forest Hill Road, Staten Island,
NY 10314, USA.

C. Luzupriaga
Department of Pediatrics, Faculty of Medicine, University of Cantabria, Avda. Herrera Oria s/n, 39011 Santander, Spain

G.R. Merriam
Research Service, American Lake VA Medical Center and Division of metabolism, Endocrinology, and Nutrition Department of Medicine, University of Washington School of Medicine, Seattle, WA 98195, USA

E. Mocchegiani
INRCA, Department of Geriatrics and Gerontological Research, Via Birarelli 8, 60100 Ancona, Italy

C. Oberti
21 California Avenue, Clementon, NJ 08021, USA

J. Oberti
21 California Avenue, Clementon, NJ 08021, USA

M.G. Obregon
Reparto di Endocrinologia, Servizio di Genetica Medica, Ospedale Pediatrico 'Bambino Gesù', IRCCS, Piazza S. Onofrio 4, 00165 Rome, Italy

M.L. Oster-Granite
Division of Biomedical Sciences, 1116 Webber Hall, University of California, Riverside, CA 92521-012l, USA

P.N. Prinz
Research Service, American Lake VA Medical Center and Sleep and Aging Research Program, Department of Psychiatry and Behavioral Sciences, University of Washington School of Medicine, Seattle, WA 98195, USA

C. Proto
Department of Pediatrics, OASI Institute (IRCCS), Via Conte Ruggero 73, 94018 Troina (EN), Italy

S. Pueschel
Child Development Center, Rhode Island Hospital, 593 Eddy Street, Providence, RI 02903, USA

L. Ragusa
Department of Pediatrics, OASI Institute (IRCCS), Via Conte Ruggero, 73, 94018 Troina (EN), Italy

P.R. Reilly
Shriver Center, 200 Trapelo Road, Waltham, MA 02254, USA

C. Romano
Department of Pediatrics, OASI Institute (IRCCS), Via Conte Ruggero, 73, 94018 Troina (EN), Italy

V.R. Sara
Department of Pathology, Karolinska Institute, Stockholm, S-104 01 Sweden

R.S. Schwartz
Harborview Medical Center and Division of Gerontology and Geriatric Medicine, Department of Medicine, University of Washington School of Medicine, Seattle, WA 98195, USA

T. Sharav
Jerusalem Child Development Center, 24 Straus Street, Jerusalem 95142, Israel

D.A. Snider
New York State Institute for Basic Research in Developmental Disabilities, 1050 Forest Hill Road, Staten Island, NY 10314, USA

V. Štrbák
Department of Clinical Immunology, Children's Hospital, Institute of Endocrinology, Slovak Academy Sciences, 833 06 Bratislava, Slovakia

M. Šustrová
Department of Clinical Immunology, Children's Hospital, Institute of Endocrinology, Slovak Academy Sciences, 833 06 Bratislava, Slovakia

C. Torrado
State University of New York Health Science Center at Brooklyn, Department of Pediatrics, 450 Clarkson Avenue, Brooklyn, NY 11203, USA

T. Tuvemo
Department of Pediatrics, University Hospital, S-751 85 Uppsala, Sweden

H.M. Wisniewski
New York State Institute for Basic Research in Developmental Disabilities, 1050 Forest Hill Road, Staten Island, NY 10314, USA

K.E. Wisniewski
New York State Institute for Basic Research in Developmental Disabilities, 1050 Forest Hill Road, Staten Island, NY 10314, USA

S. Zabransky
University of Saarland Medical Center, Division of Pediatric Endocrinology, Homburg/Saar D-6650, Germany

Foreword

Down's syndrome (DS) is the most common inherited form of mental retardation and growth retardation is one of its phenotypic characteristics, so it is natural then that clinicians would seek to treat DS with growth hormone. However, to date there is little experience in treating children with DS with human growth hormone. There are some promising findings, but the number of children treated is small and length of time of the treatments short—five years or less. Moreover, the persons who are actively involved in this work have not had an opportunity to meet and share their data and critically examine their experiences using these treatments.

The purpose of convening this International Conference on Growth Hormone Treatment in Down's Syndrome was to provide a forum (a) to give the persons studying growth hormone treatment to share their most recent data, (b) to allow the investigators and other knowledgeable persons interested in these treatments to critique the data, (c) to examine ethical and nursing issues related to these treatments, (d) to hear family perspectives about growth hormone treatment in children with DS, and (e) to discuss possible side effects of these treatments, such as increases in the risk of diabetes and leukemia.

The conference was held on 10–11 October 1992 at the Vista Hotel in New York City and was attended by 125 people, including 32 from England, Germany, Italy, Mexico, Slovakia, Spain, and Sweden. The participants heard 14 oral presentations, reviewed 13 posters, and took part in discussions ranging across the entire gamut of issues: growth retardation and neuropathology of DS, thyroid and hypothalamic dysfunction in DS, neuroimmune modulating of zinc in DS, an animal model of DS, and growth hormone as well as other treatments in DS. The participants also heard papers on and discussed ethical and nursing issues in growth hormone treatment and also heard two family perspectives on these treatments.

This book includes 13 chapters related to the invited presentations (two of the presentations were combined into a single chapter) and five shorter chapters by persons presenting posters. The discussions following invited talks were recorded and transcribed. Each speaker has been given an opportunity to clarify his or her remarks; the edited transcripts appear after the chapters when they help to illuminate important issues. The discussions at the end of the first and second days' presentations, as well as the discussion of complications, are presented as chapters in this volume. Because this was the first opportunity that many of the persons studying growth hormone treatment in

DS had to meet and to discuss their experiences, some of these discussions highlight the issues extremely well.

I would like to thank the conveners, Dr Salvador Castells and Dr Krystyna Wisniewski, for having taken the initiative to hold this conference. They planned an excellent program and brought together an outstanding group of speakers to address an important topic. I believe that this meeting will open up the discussion of growth hormone treatment in DS and that the book will be a valuable reference for physicians considering using this treatment.

I would particularly like to thank Dr Al Snider for organizing all aspects of the meeting, from developing the announcements and generating industry support to the detailed arrangements for the actual conference and the editing of this book of proceedings. I would also like to thank Dr Snider's team of Ms Marie Goepel, Ms Joan Maffei and Ms Natalie Perrone for supporting the planning and execution of the meeting so professionally. Finally, I would like to thank Ms Madeline Tinney and Ms Maffei, who expertly prepared the transcripts of the discussions and several chapters.

I would like to express our gratitude to Kabi Pharmacia AB and Eli Lilly for providing educational grants to support the conference. Pharmaceutical industry support of such conferences is essential, if we are going to educate practitioners and examine new ideas in the treatment of special populations. I would like to express our gratitude particularly to Kabi Pharmacia for publicizing the conference in Europe, and enabling several scientists to travel to New York both to present their data and to enrich the discussions with their insights.

The idea of treating short stature and growth retardation in DS with growth hormone is one we must now seriously consider, but one that we must continue to study carefully. Investigators from different clinics have shown that we can produce gains in growth in children with DS for at least five years (the longest time any of those treated have been studied to date) with no apparent complications. We need to study those treatments with protocols that will overcome recurring research design problems, such as selection bias. We must, of course, continue to monitor for possible complications (autoimmune problems, osteoporosis, diabetes mellitus and leukemia), for although diabetes and leukemia have not been observed in the studies of synthetic growth hormone treatment thus far, we must carefully watch to see if these or other possible complications emerge. Finally, we need to gather good information on other effects of these treatments; for example, do these treatments increase head circumference, and how do they affect functioning? We have some preliminary data that indicate that the head circumferences of children treated are increasing, but we need to make sure that these data are reliable and valid. We need to answer questions such as: How long do the gains in stature (and head circumference) continue? What happens to growth when the treatments are discontinued? Do the persons who are treated function better at a cognitive level and in other areas of performance? Even if one accepts the ethical principle that a child with DS has the same right to growth hormone treatment as other children of short stature, these treatments are expensive; so it is important to know how these treatments affect performance and also how the children feel about themselves.

Henryk M. Wisniewski
New York State Institute for Basic Research in Development Disabilities, Staten Island, New York

Conference Summary and Overview of the Proceedings

SALVADOR CASTELLS[1]

AND

KRYSTYNA E. WISNIEWSKI[2]

[1]Department of Pediatrics, State University of New York Health Science Center at Brooklyn, and [2]New York State Institute for Basic Research in Developmental Disabilities, Staten Island, New York, USA

The International Conference on Growth Hormone Treatment in Down's Syndrome, which was organized by the New York State Institute for Basic Research in Developmental Disabilities and was held on 10–11 October, 1992 in New York City, provided a review of the progress that is being made in the use of recombinant human growth hormone (rhGH or hGH) for treatment of short stature and microcephaly in children with Down's syndrome (DS).

The authors of this summary, who were also the conveners of the conference, were fortunate to be able to assemble many leading investigators studying DS and growth. Many outstanding papers were presented at this conference. An overview of these papers is given here, along with a summary of selected discussions following the presentations. Because this represented the first forum for the discussion and sharing of experiences with rhGH treatments in DS, the discussions, which are included in these proceedings, are most illuminating.

In Chapter 1, W. Ted Brown describes why a genetic imbalance due to trisomy 21 results in the phenotypic expression of DS. More molecular genetic studies of DS are necessary to determine how trisomy 21 causes the phenotypic expression of DS and how a range of effective treatments can be developed.

The study of growth in children with DS is important because children with trisomy 21 grow differently from normal children. Growth retardation in DS was first described in children who lived in institutions, as did many if not most children with DS until the early 1980s. Since it is known that hospitalization itself can cause growth retardation, it was not clear if the growth retardation in DS was due to trisomy 21 or the restrictive environment in which children with DS lived. The movement to close the institutions for the developmentally disabled and the desire of many parents of children with DS to keep their child at home forces us to re-examine growth retardation in these children. If growth retardation persists in children raised at home, it is not due

Growth Hormone Treatment in Down's Syndrome. Edited by S. Castells and K.E. Wisniewski

to institutional environments and psychological deprivation alone, and we need to examine what we can do to improve their growth and overall physical well-being.

Christine Cronk, Siegfried Pueschel and their collaborators have developed growth charts for children and adolescents with DS that demonstrate clearly the severity of growth retardation in these children. In Chapter 2, Dr Cronk reviews the literature on prenatal somatic growth and concludes that prenatal growth of fetuses with DS appears to be slightly reduced compared to normal children, but after birth by ages 24–36 months that growth retardation clearly manifests itself. Jesús Flórez also reported short stature in the prepubertal children with DS living at home that he and his colleagues in Spain studied. Those data are provided in Chapter 3 by Luzuriaga *et al.* in this volume. The findings of postnatal growth retardation and retardation in skeletal maturation in children with normal thyroid functions suggest to us the possibility of a partial GH deficiency, and the use of hGH for treatment.

Microcephaly and abnormal brain growth in individuals with DS are described by Dr Krystyna E. Wisniewski and her associates in Chapter 4. They stress that most children with DS have somatic growth retardation, dysmorphic features and different degrees of cognitive dysfunction, and that 80% had microcephaly. The head circumferences of the children studied were in the lower-normal range from the beginning of life, but after six months of age it was often below −2 SD. Dr Wisniewski and her associates also discuss central nervous system development, maturation and neuropathological abnormalities in DS children. They studied brains of individuals with DS as well as control brains ranging in age from birth to five years of age. Microscopic examination of the cerebral cortex showed that the subjects with DS had a 20–50% reduction of neurons since birth, mainly in the granular cell layers; and synaptic abnormalities were also present. Morphometric findings were similar in the hypothalamic nuclei (arcuate nucleus—ARC; ventromedial nucleus—VMN) of children with DS. The number of neurons per square millimeter in the hypothalamic nuclei (ARC and VMN) were significantly less in DS than in control brains.

Carmen Torrado and Salvador Castells discuss the different stimulatory tests of GH used to evaluate the basis of short stature in the DS child in Chapter 5. While it is easy to diagnose complete hGH deficiency at the level of the pituitary gland, it is difficult to measure partial deficiency due to neuroendocrine dysfunction. Children with isolated hGH deficiency are very dwarfed if not treated, and their serum concentrations of GH do not increase in response to any of the stimulatory tests. On the other hand, growth in children with DS although markedly below normal continues, suggesting a partial GH deficiency. Normal responses in serum GH concentrations to growth hormone-releasing hormone (GHRH) administration indicates normal somatotrophs and suggest a hypothalamic dysfunction. L-Dopa and clonidine have been used by Dr Torrado and her colleagues to document hypothalamic dysfunction in DS. Both L-Dopa and clonidine stimulate hypothalamic GHRH production, which stimulates release of GH into the blood. Insufficient GH peak response is the marker for dysfunction of the hypothalamic centers.

The thyroid hormone also influences growth, particularly in the first few years of life. Theresa Sharav reports on the effects of thyroid dysfunction in children with DS in Chapter 6. Dr Sharav interprets high thyrotropin (TSH) serum concentrations in DS children as evidence of immaturity of the hypothalamic–pituitary axis. The results of Torrado-Jule on hypothalamic dysfunction in the regulation of plasma GH concentrations correlate with Dr Sharav's findings of hypothalamic dysfunction in TSH regulation. Thus, it appears that the hypothalamus is dysfunctional in the regulation of the two pituitary hormones TSH and GH in children with DS, both of which are very important for growth in children. The data presented in the chapters by Maria Sustrová and Vladimir Strbák (Chapter 7), Siegfried Zabransky (Chapter 8) and Antonio Crinò and colleagues (Chapter 9) also show elevated plasma concentrations of TSH in children with DS.

There is a great need for an animal model of DS to study linear growth, as well as head and brain growth if drugs like rhGH, insulin-like growth factor 1 (IGF-1) and other growth factors are to be used in children with DS. Such a model would allow the study of the molecular action of GH and growth factors before human experimentation. Mary Lou Oster-Granite has developed a trisomy 16 mouse (Ts16) as a model system for studies relevant to children with DS, in such a way as to be able to study growth and development similar to DS. Studies of Ts16 mice have shown growth retardation and abnormal cranial–facial development. The Ts16 mouse has been found to be a very complex model to study the effects of GH in the trisomic mouse because Ts16 mice do not survive birth. As she explains in her chapter in this volume, other models are being developed that survive birth; hopefully GH treatment and its effects on linear and brain growth can be studied in these models. Oster-Granite's work is described in Chapter 10.

George Merriam's invited talk was on Treatment of Growth Hormone Deficiency with Growth Hormone-releasing Hormone. The synthesis of GHRH opens up the possibility of clinical trials of this hypothalamic hormone in the treatment of short stature in DS. His paper is presented in Chapter 11.

In Chapter 12, Zvi Laron and colleagues review the role of GH and its mediator IGF-1 in brain growth and development by studying two unique models of GH deficiency caused by hereditary molecular defects. One is caused by the hGH gene deletion and the other is Laron-type dwarfism (LTD). In both models there is little IGF-1 produced. In the first model, hGH treatment not only accelerates linear growth, but also head circumference. In LTD, daily treatment with IGF-1 markedly enhanced linear growth, but also produced a striking growth of head circumference. Thus, these two different models, which are due to genetic defects, indicate the importance of GH and

IGF-1 as regulators not only of linear growth, but also of growth in head circumference.

Salvador Castells *et al.* presented the effects of hGH in DS children with short stature for periods of treatment ranging from one to five years. His group has found that hGH treatment produces a significant increase in serum concentrations of IGF-1. They also found a persistent increase in growth velocity in all of the children treated, as well as increases in head circumference in those children treated who were below four years of age. Their data are presented in Chapter 13.

Göran Annerén and Vicki Sara each examined the role IGFs in DS in their invited talks; their data are combined and presented as Chapter 14 in this book. Included among their many important observations is that GH regulates growth from six months postnatally onward by stimulating the production of IGF-1. Six months is the critical age at which DS children start to show growth retardation and microcephaly. Drs Annerén and Sara have found IGF-1 as a growth factor promoting both linear growth and central nervous system growth. They believe that low IGF-1 may be the cause for the growth retardation and microcephaly in children with DS. They have also found an increase in serum IGF-1 concentrations during therapy with hGH. Treatment with hGH also produces an acceleration of linear growth velocity and a slight increase in the velocity of growth of head circumference.

Letizia Ragusa and colleagues presented the results of one year of hGH treatment in four children with DS. All four of the children her group studied had low serum concentrations of IGF-1 before treatment, and all four children also showed an increase in growth rate during the period of hGH treatment. Her data are presented as Chapter 15 in these proceedings.

It is not surprising that the three investigators treating DS children with hGH all found decreased serum concentrations of IGF-1 before treatment and marked increases during the period of treatment. Acceleration of linear growth and head growth also follows the increase in serum concentrations of IGF-1.

Giorgio Albertini discussed the findings of zinc deficiency in children with DS and its implications in immunological, thymic and neuroendocrine abnormalities. Chapter 16 by Fabris *et al.* details the findings of reduced turnover of zinc in subjects with DS and the physiopathological changes found in DS that might be due to zinc deficiency. Zinc supplementation corrects the zinc deficiency found before therapy and increases the number of peripheral T-lymphocyte ($CD3^+$) and T-helper/inducer cells ($CD4^+$). These changes in T-cells are reflected clinically by a marked decrease in the number of infectious episodes.

In Chapter 17, the general discussion following the Saturday presentations is given. This session was chaired by Zvi Laron of the University of Tel Aviv.

In Chapter 18, Salvador Castells and Al Snider summarize the discussion of possible complications of GH therapy. The use of rhGH in DS has produced no complications up to this point, with some children treated for up to five years. Hyperglycemia and leukemia are possible complications of particular concern. The concern about hyperglycemia is based on the finding that excess hGH, as in acromegaly, is accompanied by hyperglycemia. In the long-term therapy with hGH by Castells and his collaborators, serum blood glucose and glycohemoglobin were monitored every three months, with no significant increase in any patients. The possibility that hGH treatment in children with isolated GH deficiency might promote leukemia has been a subject of debate in the literature. That leukemia is more frequent in children with DS than in non-DS children increases the concern of treating DS with hGH. Although the number of children with DS treated with hGH by the three groups of investigators is small and the longest period of time any of them have been followed is five years, it is most encouraging to see that leukemia has not appeared in any of the cases.

In addition to the papers on the studies of hGH treatment of DS, the ethical issues, family nursing intervention and family perspectives were presented. In Chapter 19, Philip Reilly provides an ethical analysis of the use of hGH in children with DS. Among other things, he concluded that children with DS have the same moral claim to these treatments as children with short stature who are not cognitively impaired.

Arlene Hurwitz describes the nursing intervention during treatment of children with DS and short stature. The need for pediatric nurse practitioners to understand the protocols involving children in treatment with hGH and to assume the role of case managers for DS families is an important practical consideration. Her description appears in Chapter 20.

Denise Duffey and Carlos Oberti, both parents with children under treatment with hGH, presented family perspectives on the treatments. Mrs Duffey describes a reluctance of the medical community to advise parents on new protocols for the study of the effects of hGH in DS in Chapter 21. Carlos and Jean Oberti's perceptions are given in Chapter 22.

The editors/conveners would like to express their appreciation, which we are sure is shared by the other researchers at the conference, to all the parents of children with DS who are participating in clinical studies of hGH. What we have learned about the use of hGH to promote growth in height and head circumference has come about due to the strong support given to the clinical investigators by these families. As evidenced by this conference the use of hGH is an important advancement for the treatment of children with DS. We hope that the International Conference on Growth Hormone Treatment in Down's Syndrome and this book will play a role in convincing the medical community of the beneficial effects of GH treatment. We hope that it will also encourage others to explore other biochemical replacements of products of damaged genes in DS and ultimately gene replacement itself.

Part I

GENETICS, GROWTH RETARDATION, MICROCEPHALY, BRAIN GROWTH AND NEUROENDOCRINE DYSFUNCTION OF GROWTH HORMONE AND THYROTROPIN

1 Molecular Genetics of Down's Syndrome

W. TED BROWN
Department of Human Genetics, New York State Institute for Basic Research in Developmental Disabilities, New York, USA

It is a generally accepted theory that the features of DS are the result of a genetic imbalance due to the underlying presence of an extra chromosome 21. Having 50% more of the genes from chromosome 21 present implies that 50% more of these gene products will be active in the individual cells from a person with DS. The goal of molecular genetic studies of DS is to identify the genes which are encoded by chromosome 21 in order to understand how the syndrome results from this genetic imbalance.

Currently, it is estimated that the total number of human genes is between 50 000 and 100 000 (1). Further, it appears that chromosome 21 is a fairly typical chromosome in terms of the density of genes per chromosome. Since chromosome 21 constitutes approximately 1.5% of the total amount of the human genetic material, we can estimate that there are between 750 and 1500 genes encoded by chromosome 21. To date, there have been approximately 40 chromosome 21 genes mapped, which represent perhaps 2.5–5% of the genes encoded by the chromosome. Several specific diseases and related genes have been mapped to genes on chromosome 21, including acute myeloid leukemia, leukocyte adhesion deficiency, hemolytic anemia due to phosphofructokinase deficiency, and homocystinuria (1). Recently mapped genes on chromosome 21 with mutations known to cause genetic diseases include the amyloid precursor protein (APP) related to some forms of familial Alzheimer's disease (2), a gene for progressive myoclonus epilepsy (3), and a gene for some cases of familial amyotrophic lateral sclerosis (4).

Recent progress has succeeded in completing a continuous array of overlapping clones covering the entire long arm of chromosome 21 from human yeast artificial chromosome libraries using sequence-tagged sites as landmarks specifically detected by polymerase chain reaction (5). A number of novel approaches, such as 'exon trapping' (6), are being developed to isolate and identify coding gene sequences. Now the complete set of gene coding sequences on chromosome 21 needs to be identified and mapped, and their functions determined.

As humans, we are each carriers for many recessive diseases. Carrier status is usually the result of the lack of expression of one of the two genes we have

Growth Hormone Treatment in Down's Syndrome. Edited by S. Castells and K.E. Wisniewski

inherited from each parent for each genetic locus. Carriers of most recessive diseases, such as cystic fibrosis and Tay-Sachs' disease, have no apparent ill effects even though they have only 50% of the normal amount of gene product present. Therefore, we can assume that having 50% more of many genes would produce no ill effects. Structural components, rate-limiting enzymes and cell surface receptors are likely gene candidates for producing the phenotype of DS.

Among the approximately 40 genes mapped, so far, to chromosome 21 are several growth regulating genes known as oncogenes, including acute myeloid leukemia oncogene 1, oncogene ERG and oncogene ETS-2. The function of these genes and their relation to the phenotype of DS are not yet understood completely. However, it is known that some oncogenes can stimulate or act to inhibit growth and thus three copies of such genes might contribute to the phenotype. Perhaps the growth stimulatory effect of growth hormone in DS may be in part due to overcoming the effect of the excessive expression of a growth-inhibiting oncogene. Once the specific genes involved in causing the DS phenotype are identified, new gene therapies may be devised to overcome their specific or growth-inhibiting effects.

Various approaches, principally the study of individuals with partial trisomy or partial monosomy 21, are being undertaken to identify the genes important to DS and to create a 'phenotypic' map of chromosome 21 (7, 8). We anticipate with the impetus of the Human Genome Project and with research efforts devoted to identifying the specific genes on chromosome 21 which contribute to the DS phenotype, that the future will be bright. We will begin soon to understand the molecular basis for this important genetic syndrome and determine how treatments may be scientifically engineered to improve the underlying genetic imbalance.

REFERENCES

1. McKusick VA. *Mendelian Inheritance in Man: Catalogues of Autosomal Dominant, Autosomal Recessive, and X-linked Phenotypes*, 10th edn, Johns Hopkins University Press, Baltimore, MD, 1992; pp. XXXIX–CCXIV.
2. Goate A, Chartier-Harlin MC, Mullan M, Brown J, Crawford F, Fidani L and 15 collaborators. Segregation of a missense mutation in the amyloid precursor protein gene with familial Alzheimer's disease. *Nature* 1991; **349:** 704–706.
3. Lehesjoki A-E, Koskiniemi M, Sistonen, P, Miao J, Hastbacka J, Norio R, de la Chapelle A. Localization of a gene for progressive myoclonus epilepsy to chromosome 21q22. *Proc Natl Acad Sci USA*, 1991; **88:** 3696–3699.
4. Siddique T, Figlewicz DA, Pericak-Vance MA, Haines JL, Rouleau G, Jeffers AJ, and collaborators. Linkage of a gene causing familial amyotrophic lateral sclerosis to chromosome 21 and evidence of genetic-locus heterogeneity. *N Engl J Med* 1991; **324:** 1281–1382.

5. Chumakov I, Rigault P, Guilou S, Ougen P, Billaut A, Guasconi G and 30 collaborators. Continuum of overlapping clones spanning the entire human chromosome 21q. *Nature* 1992; **359:** 380–387.
6. Duyk GM, Kim SW, Myers RM, Cox DR. Exon trapping: a genetic screen to identify candidate transcribed sequences in cloned mammalian genomic DNA. *Proc Natl Acad Sci USA* 1990; **87:** 8995–8999.
7. Korenberg JR, Kawashima H, Pulst SM, Alen L, Magenis E, Epstein CJ. Down's syndrome: toward a molecular definition of the phenotype. *Am J Med Genet Suppl* 1990; **7:** 91–97.
8. Epstein CJ, Korenberg JR, Anneren G, Antonarakis SE and 18 collaborators. Protocols to establish genotype–phenotype correlations in Down's syndrome. *Am J Hum Genet* 1991; **49:** 207–235.

2 Growth Retardation in Down's Syndrome

CHRISTINE E. CRONK
Department of Anthropology, Southern Illinois University, Carbondale, Illinois, USA

INTRODUCTION

Abnormal growth in children with Down's syndrome (DS) has been thoroughly and repeatedly documented since the earliest studies of this condition. While the available research has allowed a clear picture of growth dynamics at some developmental stages (for example, infancy), growth characteristics of other periods in the life (for example, childhood and late adolescence) are less clearly established. In addition, body composition changes are poorly understood. Finally, any research concerning treatment effects on growth with children with DS must consider methodological issues which affect conclusions in samples of this kind.

Much of the information presented here has been summarized in the recently published book edited by Pueschel and Pueschel (1). The following questions will be addressed here:

1. What are the characteristics of linear growth (as measured by body, segment and limb lengths) and body composition in children with DS? What factors affect growth?
2. How does variability and continuity of growth in children with DS differ from that for normals?
3. What are some methodological issues of concern in the study of growth in children with DS?

PRENATAL GROWTH

Studies prior to the 1980s relied primarily on size at birth for information about prenatal growth. In general, these studies agreed that both weight and length at birth are slightly reduced relative to normal. Smith and McKeown (2) found a difference of -0.3 kg from the average weight of their sample compared to normal controls. The difference was partly confounded with gestational age. However, weight for infants of 38 weeks gestational age and less

Growth Hormone Treatment in Down's Syndrome. Edited by S. Castells and K.E. Wisniewski

was also less than the normal mean. These findings were corroborated in later studies (3–5). Ershow (6) found a difference in birth size between black and white infants with DS using data from the perinatal collaborative study, which was ultimately attributed to differences between these two groups in maternal cigarette consumption and maternal weight at delivery.

Pueschel *et al.* (7) controlled for familial genetic effects on birth size using data on sibling birth weight. Mean birth weight for infants with DS (3.07 kg) was significantly less than that for their siblings (3.52 kg), adjusting for gestational age, birth order and maternal age. These investigators concluded that growth retardation was associated with the syndrome and independent of the factors controlled for in their study.

Results of studies assessing birth length of infants with DS have been contradictory (8–11). Using data collected in the study of Pueschel *et al.* (7), Cronk and Pueschel (12) found slightly, though significantly, shorter birth lengths in newborns with DS than those for a normal control group (48.9 cm versus 50 cm). However, this difference was not significant after gestational age was controlled.

Using serial ultrasound measurements, Kurjak and Kirkinen (13) followed eight fetuses with trisomy 21 from 16 to 38 weeks of gestation. All fetuses showed normal biparietal diameters, but abdominal circumference was slightly reduced compared to normal. It is noteworthy that the brachycephalic conformation of the head in individuals with DS, if manifest prenatally, would lead to normal biparietal diameter. Other ultrasound studies have indicated that fetal femur length is slightly shortened compared to reference values (14). In a study of abortuses with DS (15), subtle growth retardation in all long bones was demonstrated, with retardation greater in bones of the upper extremities. Because of small sample size, the time of onset of reduced long bone growth could not be determined.

In summary, prenatal growth of fetuses with DS appears to be slightly reduced compared to normal. Birth weight is slightly reduced compared to normal even controlling for gestational age, parity, maternal age, and family endowment for birth size. Age of onset of this weight deficit is uncertain. Growth in body and limb length during fetal life also appears to be reduced. Again, onset age for linear growth deficiency has not been established.

GROWTH DURING INFANCY

Postnatal growth retardation has been a clearly identified feature of DS from the earliest studies. Most investigations have agreed that by adulthood stature is reduced by an average of about 2 SD (i.e. is less than fifth percentile), and many have documented that the severest deficit occurs during infancy or early childhood (8, 10–11, 16–19). In general, these studies have established deficits of nearly 2 SD (less than fifth percentile) by two years of age, with mean deficits as large as 4 SD between two years and adulthood. Roche (8) found no difference in the degree of deficit between the sexes in his sample and a greater range of variation in size for DS than for normals.

A longitudinal study of infants with DS was conducted at the Children's Hospital of Boston during the 1970s (12, 20–21). This study evaluated approximately 100 home-reared infants followed at three- and then six-monthly intervals from birth until three years and then yearly thereafter. Findings from this study form the basis for many of the statements below.

Figures 1 and 2 graph growth for height from birth to three years compared with values for the National Center for Health Statistics sample (22). Birth length and weight were slightly reduced from normals (⩽1 cm in each sex). Progressive decline in average recumbent length relative to normal controls was apparent. At three months mean values for DS and normals were about 2 cm apart (−0.9 SD or at about 15th percentile). By three years this difference in the means was 3.5 cm (−2.5 SD or at about first percentile). Variability was broader for children with DS (larger SD bars on Fig. 1 for the children with DS). Boys with DS were more variable than normal from birth to 36 months, whereas girls had variability similar to normals starting at nine months of age.

Six-month velocity of growth rate from birth to 36 months was reduced by about 1 SD (around 10th to 25th percentile). For the full 36 months the average child with DS grew 38.2 cm (SD 4.05 cm, range 28.2–47.5 cm), whereas the average normal child grows 46 cm during this same period.

There was substantial inter-individual variation in growth deficiency for this period. About 70% demonstrated deficient growth in all six-month intervals

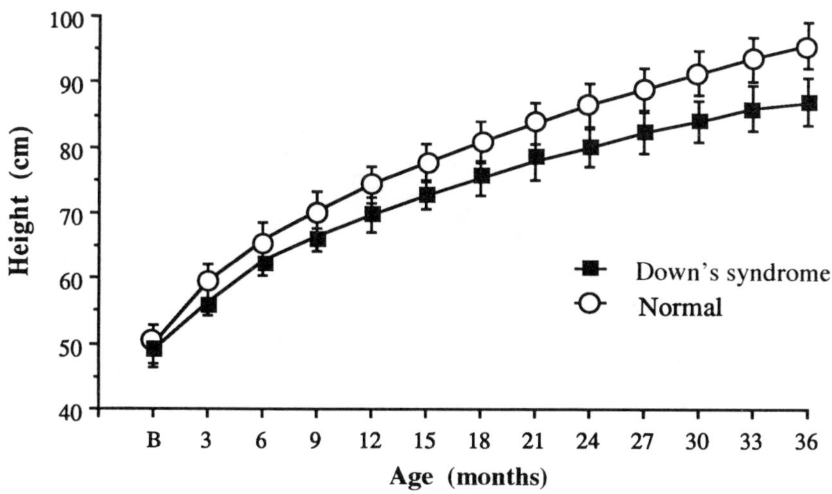

Figure 1. Mean height (recumbent length) in centimeters from birth to 36 months for girls from the Boston Children's Hospital sample compared with age- and sex-specific mean (±1 SD) for National Center for Health Statistics (22)

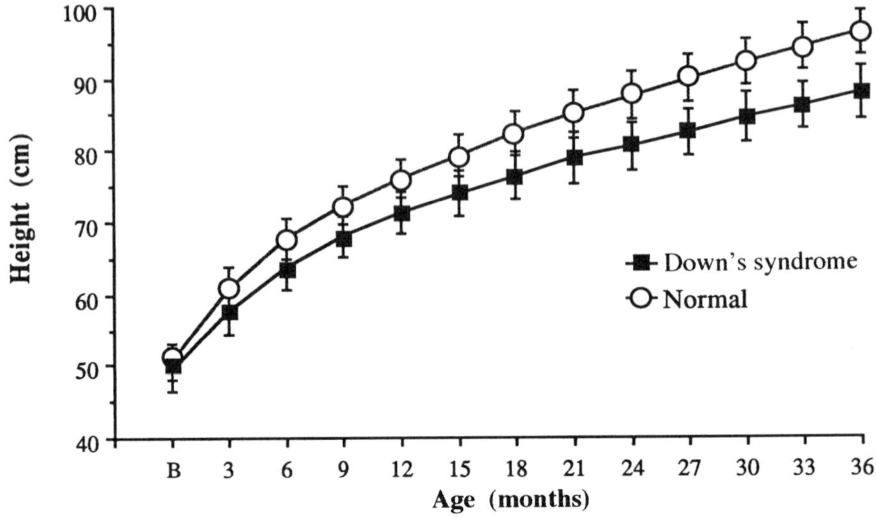

Figure 2. Mean height (recumbent length) in centimeters from birth to 36 months for boys from the Boston Children's Hospital sample compared with age- and sex-specific mean (±1 SD) for National Center for Health Statistics (22)

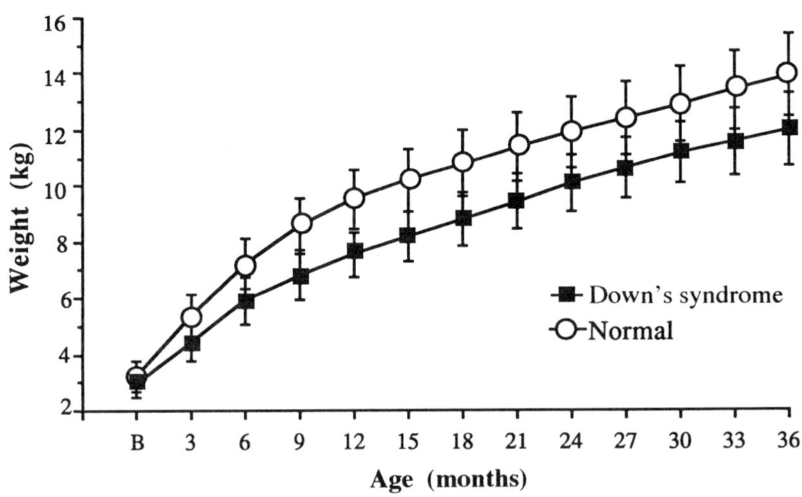

Figure 3. Mean weight in kilograms from birth to 36 months for girls from the Boston Children's Hospital sample compared with age- and sex-specific mean (±1 SD) for National Center for Health Statistics (22)

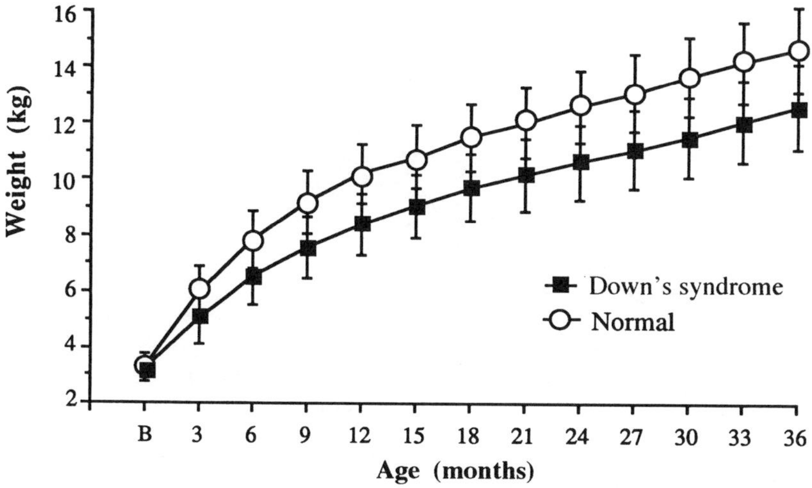

Figure 4. Mean weight in kilograms from birth to 36 months for boys from the Boston Children's Hospital sample compared with age- and sex-specific mean (±1 SD) for National Center for Health Statistics (22)

from birth to 36 months of age. These infants were less than fifth percentile by the end of infancy. About 30% grew intermittently at a normal velocity. These periods were punctuated with intervals of deficient growth. Only about 10% had normal growth velocity throughout infancy and early childhood. Frequently, growth deficiency was so great that periods of 'no growth' lasting three to six months were common.

Mean values for weight from birth to 36 months from the Children's Hospital of Boston sample compared to the National Center for Health Statistics reference data are graphed in Figs 3 and 4. Values for weight and weight velocity were similar to those for length. Differences between DS and normals at three months were about 1 kg (−1 SD or about 10th percentile). By three years the difference was 3 kg in each sex (−2 SD or less than third percentile). Velocity of weight gain was reduced from 6 and 18 months when weight gain was 22% less than for normal children. Between 18 and 36 months velocity was not different from normal.

GROWTH AFTER INFANCY

Cross-sectional data for childhood and adolescence growth are readily available. However, no comprehensive serial studies have been carried out evaluating growth during these periods. Figures 5 and 6 present data on height at 4–18 years drawn from four different studies compared to the National Center for Health Statistics (NCHS) data (22). From early childhood until about

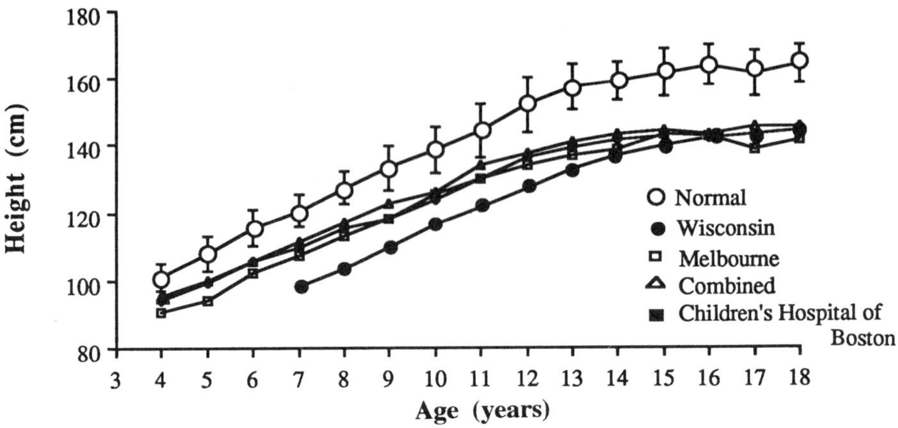

Figure 5. Height in centimeters for 4–18 years for four samples of girls with Down's syndrome: Boston Children's Hospital (12, 20, 24), Wisconsin (19), Combined (24), Melbourne (8) and normal (±1 SD) (22)

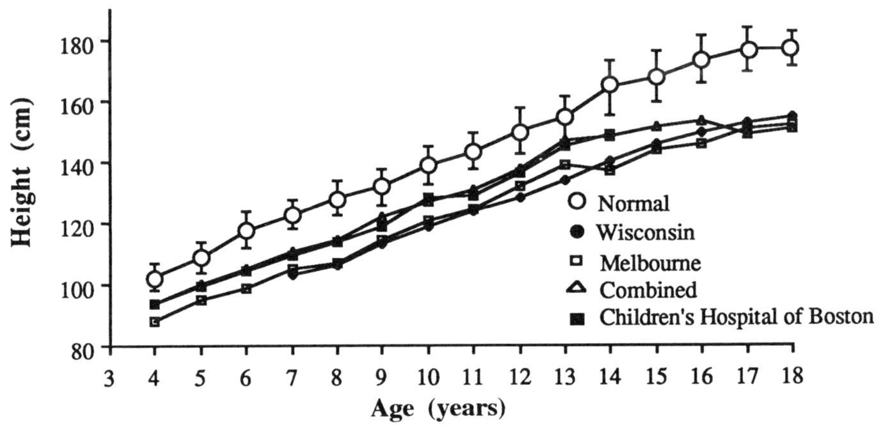

Figure 6. Height in centimeters for 4–18 years for four samples of boys with Down's syndrome: Boston Children's Hospital (12, 20, 24), Wisconsin (19), Combined (24), Melbourne (8) and normal (±1 SD) (22)

11 or 12 years of age for girls, and 15 or 16 years of age in boys, the difference between children with DS and normal children is similar. After these ages, the distance between the normal and DS curves increases.

CHILDHOOD GROWTH

Several studies of institutionalized children have found statural deficits of between 2 and 4 SD for all ages from 3 to 10 years (8–11, 18). Ikeda *et al.*

(23) assessed home-reared Japanese children with DS from infancy to four years. Scores declined until about 30 months (about 3 SD below the control group mean) and then stabilized through early childhood (48 months). Six-month velocities after 36 months were usually less than normal. While these findings are generally similar to those from other studies, the pattern of early childhood growth velocities indicates that growth continues to be deficient through early childhood.

Ershow (6) found size at age seven years to be reduced between 2 and 3 SD for height and around 0.5 SD for weight, though at 12 months these children had mean values reduced by only about 1.5 SD. Longitudinal data collected by Rarick and Seefeldt (19) showed that at age seven years height was reduced by 3–4 SD but did not change between 7 and 10 years of age. Taken together, these studies suggest a continuing, though less marked, growth retardation during childhood as compared with infancy.

Mixed longitudinal data on home-reared children gathered from three US east coast centers (24) indicated that between 2 and about 11.5 years girls with DS were close to 2 SD below the NCHS mean, with minor fluctuations (−1.5 to −2.49 SD). Boys were about 2 SD less than the NCHS mean (−2.3 to −3 SD). Weight deficits were less, ranging from 1.2 to 0 SD. Pseudovelocities (i.e. rate of gain computed by subtracting an earlier from a later mean value) were between the 3rd and 25th percentile for males, and between the 3rd and 50th percentile in females. Pseudovelocities for weight were between 25th and 50th centiles.

A subsample of children from the Children's Hospital of Boston study was

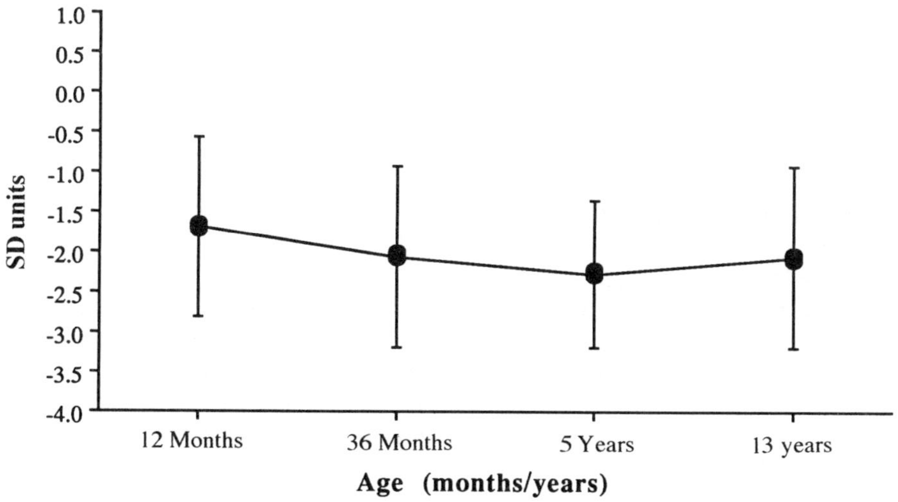

Figure 7. Mean (±1 SD) z score for children from the Boston Children's Hospital sample at one, three, five and 13 years

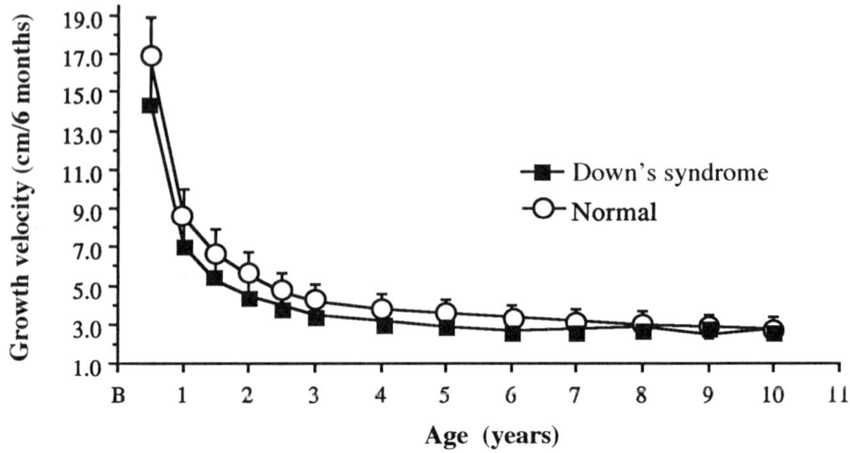

Figure 8. Mean growth velocity (centimeters per six months) from birth to 10 years for children from the Boston Children's Hospital sample (12, 20, 24) contrasted with mean (±1 SD) for normals (26)

assessed intermittently during childhood and again at between 12 and 15 years. Figure 7 plots mean *z* scores for height for these children at one, three, five and 13 years. Mean *z* scores were similar beginning at three years (around −2.00 SD). Mean six-month velocity (sexes combined) from birth to 10 years for this group was compared to reference data from the Fels Institute study (25) (Fig. 8). Velocity of growth was less than normal at least until eight years of age. These findings reinforce the observation that growth through mid-childhood is slow. This is in turn corroborated by Rarick *et al.* (26), who found that the prepubertal component of growth curves fitted using the double logistic function contributed importantly to the final height of children with DS.

GROWTH AT ADOLESCENCE

During adolescence, height is reduced by 2–4 SD below the normal mean for all of the samples plotted on Figs 5 and 6. Values for the two institutionalized samples (8, 19) were significantly lower than that for home-reared children (24) with differences of between 7 and 12 cm. At the end of adolescence, however, all samples appear to have size reduced by −3.5 and −4 SD. Home-reared children may thus have a nutritional and/or emotional advantage throughout the childhood growing period, such that final height is reached earlier. Pseudovelocities were in the lower percentiles (25th to 50th percentile) for normal in each sex.

Roche (8) found that 90% of the children in his sample experienced an adolescent growth spurt (annual height increment ⩾4 cm). Peak velocities ranged from 5 to about 13 cm per year (similar to normal). These spurts occurred at ages similar to those of normals. Final height was reached significantly earlier than in normals (15.5 years, range 13.9–18 years in boys; 14.3 years, range 12.1–16 years in girls). At the time of growth cessation, bone ages were not fully mature.

Rarick and Seefeldt (19) found growth less than normal in their sample (ages 10–18 years). Growth of boys in their sample was less than normal adolescents, whereas in girls no difference was apparent. When they subdivided the full period into two shorter ones (10–14 years and 14–18 years) girls appeared to have significantly greater growth than normal girls in later puberty. This suggested a later than normal occurrence of the pubertal growth spurt in girls with DS.

Peak annual height velocity for boys with DS was 7.83 cm (compared with 9.14 cm for normal boys), and for girls was 6.61 (compared with 7.83 cm for normal girls). The age of peak growth was alike in children with DS (11.8 years for girls and 13.79 years for boys) and normals (12 years for girls and 14 years for boys). However, analyses fitting double logistic curves to the DS and control data suggested that the growth spurt for both boys and girls with DS occurred a year later than normals (26).

Timing of maturation and the appearance of secondary sex characteristics in children with DS have only rarely been investigated. Rundle (27) found that menarcheal age, breast and pubic hair stages were only slightly delayed in girls with DS. Pueschel (28) reported essentially normal values for menstrual characteristics (average age of menarche was 12 years 6 months, menstrual period intervals 27 days and menstrual duration four days). He also found that physiological symptoms (headache, mood swings, abdominal bloating) similar to those in normal occurred in about 70% of his sample. Buday (29) reported an average menarcheal age of 13.45 years for a group of institutionalized girls in Hungary compared to a control group mean of 13 years.

In the study of Pueschel *et al.* (30), timing and appearance of most secondary sex characteristics in males with DS were similar to normal. However, axillary and facial hair (particularly hair on the upper lip) developed later than normal. Testicular volume and penile length were about the same as normal controls. These investigators also found that changes of hormone levels including follicle-stimulating hormone, lutenizing hormone and testosterone were similar to those of normal adolescents.

GROWTH OF BODY SEGMENTS

A few studies have assessed growth of lateral and segmental dimensions of children with DS. Both Thelander and Pryor (31) and Rundle (27) evaluated retardation of sitting height and stature, and found greater deficiency in lower body than in upper body growth. Rarick *et al.* (32) investigated tibial length from 7 to 18 years. Mean values were found to be those typical for children four years younger.

Segmental disproportion was most clearly demonstrated in studies of Ikeda *et al.* (23) and Rarick and Seefeldt (19). Ikeda *et al.* (23) found that the upper limb length/stature and lower limb length/stature ratios of children with DS differed from normal at most ages from 3 to 48 months, indicating disproportionately short limbs. In contrast, the intermembral index (upper limb/lower limb) was similar to normal. Rarick and Seefeldt (19) found the ratio of sitting height to stature to be significantly different from normal in their sample of 7–18-year-olds, demonstrating the relatively shorter leg lengths of these children at all ages. Sex differences and age changes in the ratio of sitting height to stature were not, however, different from normal in this group, suggesting a similar sex-specific schedule for relative growth in the components of linear growth.

SKELETAL MATURATION

Maturation as measured by bone age using hand and wrist radiographs is used as a direct measure of biological age. Relative to the Greulich and Pyle (33) standards, children from the Children's Hospital of Boston sample had mean bone ages of 17 ± 6 months at 24 months and 28 ± 7 months at 36 months. These values were reduced by about 1 and 2 SD below the mean respectively. The variability for bone ages was similar to that for normal children. When analyzed categorically, about 75% had retarded, 9% had advanced, and 16% had normal bone ages at 24 or 36 months. Most children who had bone age evaluations at both ages showed enhanced skeletal retardation at the later assessment.

After infancy, Roche (34) observed that skeletal maturation levels increased with relation to normal from about 4 to about 10 years of age. During this time, children with DS have bone age delays of 1.5–2 years (about −1 SD below the normal mean). After 10 years, average values remained similar until the end of the growing period. Observations of Rarick *et al.* (32) observed marked variability among children, with 67% of boys and 72% of girls more than 2 SD below the normal mean for age.

MEASURES OF BODY FATNESS

Most studies have analyzed body fatness using weight in relation to height. Brousseau and Brainerd (9) and Benda (10) noted a lack of uniformity in weight/height relationships in their studies. Data on three samples—two institutionalized (the Melbourne sample originally reported by Roche (8), and the Wisconsin sample originally reported by Rarick and Seefeldt (19) and one home-reared sample (the Children's Hospital of Boston sample (12, 20, 24)—were analyzed using sex- and age-specific medians for body mass index (weight/height^2) compared with normal children (35). Beginning at about two years of age children with DS in each group had median values between 75th and 90th percentiles for normal at some ages between 2 and 12 years. After 12 years, median values for the two institutionalized samples were similar to those for the controls, possibly owing to the difference in peak height velocity for the two groups.

Weight for 5 cm height intervals by sex for these same groups was compared to control data (36). Values were clearly above those for normal children beginning at about the 100 cm interval for height (around four to six years of age). Values remained above the normal mean for all the remaining intervals. Institutionalized children had a greater propensity for greater weight for height than those who were home-reared.

In a recent report on institutionalized children from Hungary, Buday (29) reported distributions for six skinfold sites in individuals of four years to adulthood. Children with DS showed higher values only at the subscapular site. The differences increased with age, particularly in girls. The other sites (triceps, biceps, suprailiac and abdominal) did not differ from normals.

FACTORS AFFECTING GROWTH

Most findings discussed under this heading are based on the Children's Hospital of Boston sample. Length and weight for infants born before 38 weeks of gestational age (range 31–38, $n = 22$) were significantly less than children with gestational age $\geqslant$ 38 weeks at birth and at three months. However, the difference disappeared by six months of age, thus indicating that, as in normal prematurely born infants, preterm infants with DS grow faster in very early infancy.

Children who have DS and a significant cardiac defect grew and gained weight more slowly during infancy. The absolute size differences between children with moderate and severe congenital heart disease and those with mild or no congenital heart disease were about 1 cm near birth, but increased to around 2.5 cm by three years of age. Growth velocity differences were small but consistent (0.3–0.6 cm per six months) until about 30 months of age, after which velocity in both groups was similar. Average weight of children with significant congenital heart disease was also lower than in children with mild or no cardiac problems, though weight velocity in the two groups did not differ consistently across the age interval. Ershow (6) observed weight deficits in children from the perinatal collaborative sample.

A small group of parents and siblings from the Children's Hospital of Boston sample were measured, and standard scores for their heights compared with height at age three years for children with DS. Correlations were low and statistically non-significant for parent size. There was a statistically significant but modest association with sibling height ($r = 0.40$, $p < 0.05$, $n = 34$), supporting the contention that genetic endowment for size is weakly manifest in growth of children with DS.

VARIABILITY AND CONTINUITY OF GROWTH

Studies of size in infancy have shown that variance of recumbent length for boys with DS is significantly greater than normals at all ages from birth to 36 months, while for girls greater variability is apparent only at birth and three months (2) (see above and Figs 1 and 2). For six-month growth velocity, variances for both sexes were similar to those for reference values until the end of infancy. For all ages from 4 to 10 years, variances were significantly greater in DS for each sex as compared with a normal reference group (25). This greater range of variation may be due to continued greater inter-individual variation in growth velocity or to the expected excess in measurement error (37).

Cronk and Reed (38) analyzed serial data from the Children's Hospital of Boston sample using longitudinal principal components analysis (LPC). This technique allowed characterization of patterns of growth compared with an internal standard. The average growth curve for height (Figs 1 and 2) represents the growth deficiency in these children as progressive and smooth, with fall-off from within the normal range to less than fifth centile by the end of infancy. LPC analysis demonstrated a number of patterns for this fall-off. Instances of progressive centile shift during infancy/early childhood were common. However, many children displayed patterns of shifts both upwards and downwards around the DS mean curve. Cronk and Reed hypothesized that the sporadic quality of growth, manifest in greater inter-individual and intra-individual variability in velocity, was evidence of poor canalization (39), i.e. decreased endogenous control over the growth process. However, some studies (40, 41) have shown that a similar pattern of sporadic growth is evident in normal children over very short periods (one week to one month). It is possible that growth in DS including prolonged periods of growth quiescence, is simply a slowed-down version of the normal process, as has been suggested by Cicchetti and Sroufe (42) for cognitive and emotional development in these children.

Simple Pearson correlation analyses also document instability in growth of children with DS. Amongst normal children, values of around 0.9 for contiguous ages and between 0.75 and 0.80 for more distant ages are characteristic (43). For children with DS, major departures from this pattern were evident for correlations of ages in infancy with those in childhood (ranging from 0.40 to 0.80). In contrast, correlations within early childhood (three to six years) were all around 0.80, and similar to those for normal children. These findings force the interpretation that infancy growth is particularly destabilized in children with DS.

Correlation analysis can also be used to predict relative size of the child at later ages. Using standard scores computed from the NCHS data (22), correlations between height at age one, three and six years with height at about age 13 years were evaluated. Figure 9 plots correlations for children with DS from the Boston Children's Hospital. Children without or with only mild congenital heart disease (CHD) are plotted separately from those with moderate or severe heart disease. Correlations were lower for children with CHD than those without, and were lower between *z* score at 60 months and 13 years than for the two earlier ages. Correlations for the healthy children were low for one and three years, but rose to about 0.70 by 60 months. These correlations are lower than those observed for normal children.

Examples of instability in size from early to later childhood are illustrated in Figs 10 and 11. Figure 10 shows a child who failed to thrive during infancy until he experienced spontaneous closure of a heart defect. Figure 11 illustrates a girl free of heart disease who was near fifth centile up to age six years. At her 13-year follow-up, her height was reduced by about 4 SD. In contrast to these examples, Fig. 12 shows a boy with stable growth throughout his life. At age three his *z* score was +0.56 SD (about 60th percentile) and remained at this level at his 13-year follow-up.

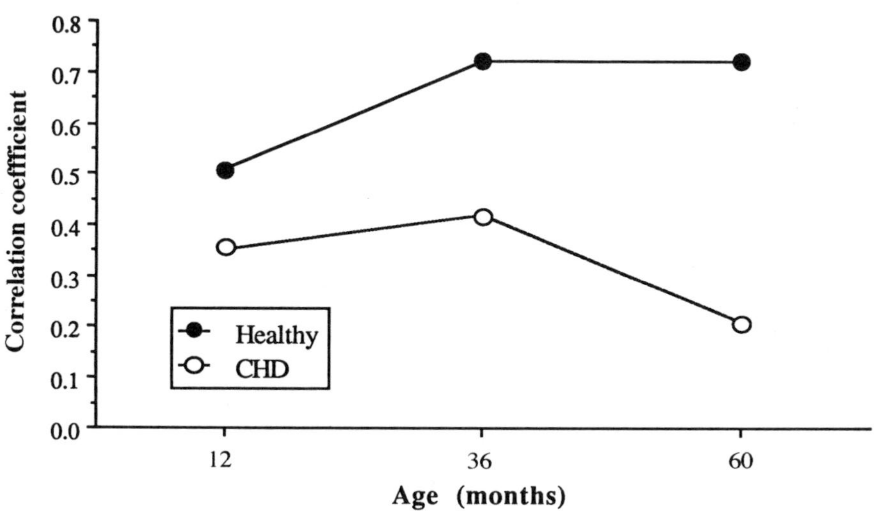

Figure 9. Pearson correlation coefficients of height/recumbent length at one, three and five years with height at 13 years for children without or with mild coronary heart disease (CHD), and those with moderate or severe heart disease.

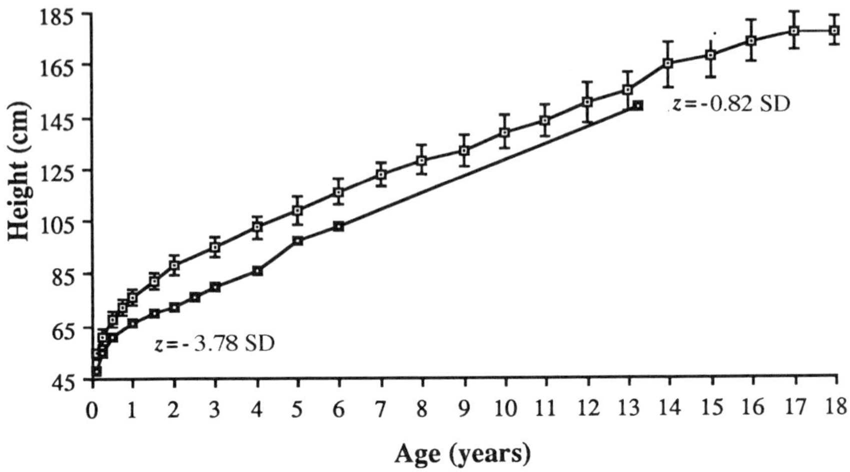

Figure 10. Height of a boy with Down's syndrome with heart disease plotted against mean (±1 SD) for normal children (22)

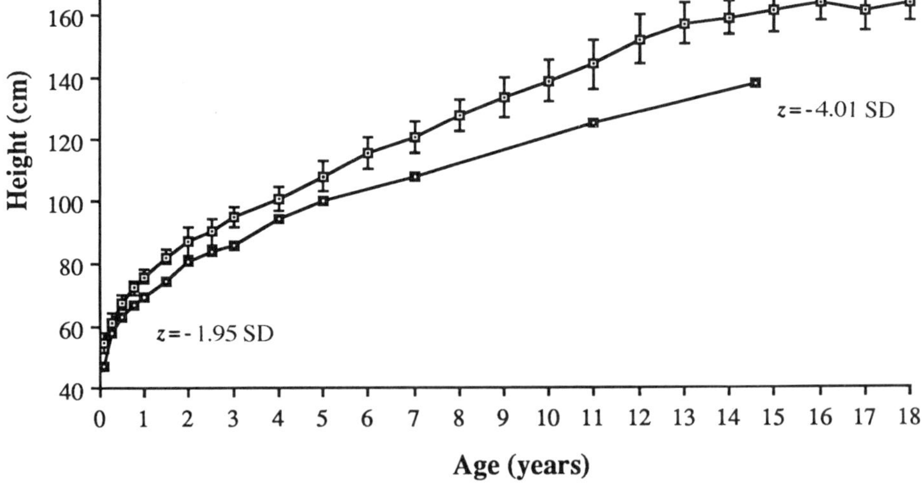

Figure 11. Height of a girl with Down's syndrome plotted against mean (±1 SD) for normal children (22)

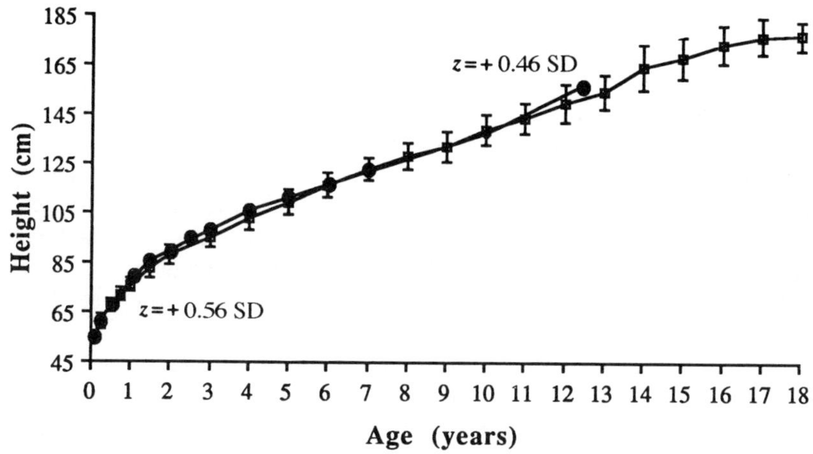

Figure 12. Height of a boy with Down's syndrome plotted against mean (±1 SD) for normal children (22)

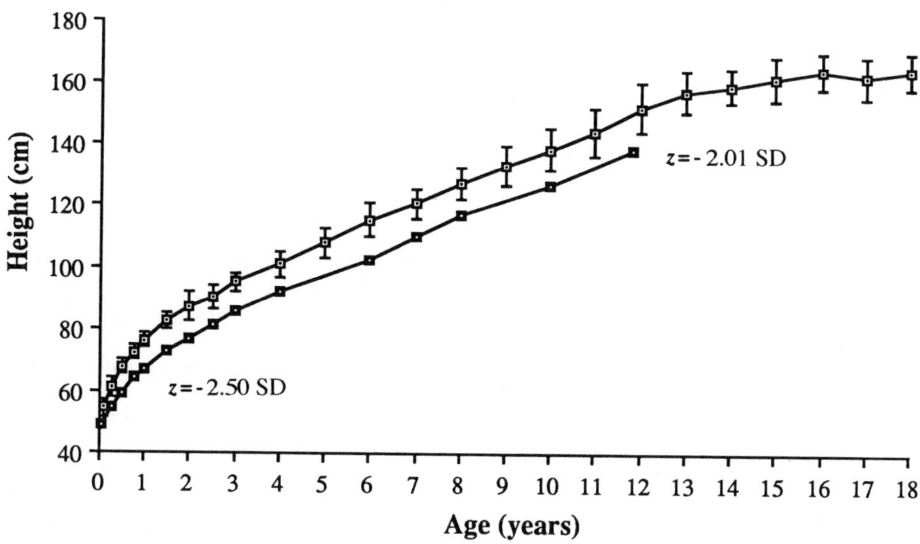

Figure 13. Growth of a girl with low stable *z* scores relative to the normal mean (±1 SD)

METHODOLOGICAL CONSIDERATIONS

Two key methodological issues should be of concern to those investigating growth of children with DS. First, the lack of generalizability of work based on small sample sizes is enhanced in these children because of their greater inter- as well as intra-individual variation. Particularly during infancy, some children experience periods of normal growth, interspersed with periods of slow or zero growth. Study designs should therefore sample longer time intervals, use methods of data analysis which are sensitive (e.g. growth velocity instead of just absolute height), and evaluate these data using age- and sex-specific reference data (e.g. the velocity standards from the Fels Research Institute (25).

Secondly, measurement error variation should be quantified so that its potential effect on the results can be evaluated. Measurement error is the unavoidable noise from technique, instruments, circumstances, and the child. Growth velocity, the more sensitive measure of treatment effects on growth, is more confounded with measurement error. It has two associated errors: that owing to noise in the first and that owing to noise in the later measurement. In clinical circumstances with a not fully cooperative child, this doubled error can be larger than growth expected even during a six-month interval. Measurement error can be quantified for a study series, or for an individual, by computing the technical error of measurement (44). Having an estimate of noise then allows the investigator to judge treatment effects relative to it.

REFERENCES

1. Pueschel SM, Pueschel JK. *Biomedical Concerns of Persons with Down's Syndrome*, Paul H. Brookes, Baltimore, 1992; pp. 19–38.
2. Smith A, McKeown T. Prenatal growth of mongoloids. *Arch Dis Childh* 1956; **30:** 448–456.
3. Kucera J, Dolezalova V. Prenatal development of malformed fetuses at 28–42 weeks of gestational age. *Biol Neonat* 1973; **22:** 319–324.
4. Papp Z, Adam S, Szabo Z. Prenatal growth in Down's syndrome *Orvosi Hetilap* 1976; **117:** 277–282.
5. Marmol JG, Scriggins AL, Vollman RF. Mothers of mongoloid infants in the Collaborative Project. *Am J Obstet Gynecol* 1969; **104:** 533–543.
6. Ershow AG. Growth in black and white children with Down's syndrome. *Am J Ment Defic* 1986; **90:** 507–512.
7. Pueschel SM, Rothman KJ, Ogilby JD. Birth weight of children with Down's syndrome. *Am J Ment Defic* 1976; **80:** 442–445.
8. Roche AF. The stature of mongols. *J Ment Defic Res* 1965; **9:** 131–45.
9. Brousseau K, Brainerd HG. *Mongolism*. William & Wilkins, Baltimore, 1928.
10. Benda CE. *Mongolism and Cretinism*. Grune & Stratton, New York, 1946.
11. Oster J. *Mongolism*. Danish Science Press, Copenhagen, 1955; **61:** pp. 564–68.
12. Cronk CE, Pueschel SM. Anthropometric studies. In *The Young Child with*

Down's Syndrome, edited by Pueschel SM, Human Science Press, New York, 1984; pp. 105–142.
13. Kurjak A, Kirkinen P. Ultrasonic growth pattern of fetuses with chromosomal aberrations. *Acta Obstet Gynecol Scand* 1982; **61:** 223–225.
14. Benacerraf BR, Gelman R, Frigoletto FD. Sonographic identification of second trimester fetuses with Down's syndrome. *N Engl J Med* 1987; **317:** 1371–1376.
15. Fitzsimmons J, Droste S, Shepard TH, Pascoe-Mason J, Chinn A, Mack LA. Long-bone growth in fetuses with Down's syndrome. *Am J Obstet Gynecol* 1989; **161:** 1174–1177.
16. Benda CE. Studies in mongolism: growth and physical development. *Arch Neurol Psychiatry* 1939; **41:** 83–95.
17. Hutton G. The physical development of Mongols. *Arch Dis Childh* 1958; **34:** 46–50.
18. Gustavson K-H. *Down's Syndrome. A Clinical and Cytogenetical Investigation* Alonquist Wiksell, Uppsala, 1964.
19. Rarick GL, Seefeldt V. Observations from longitudinal data on growth in stature and sitting height of children with Down's syndrome *J Ment Defic Res* 1974; **18:** 63–78.
20. Cronk CE. Growth of children with Down's syndrome: birth to age 3 years. *Pediatrics* 1978; **61:** 564–568.
22. Pueschel SM. The study population. In *The Young Child with Down' s Syndrome*, edited by Pueschel SM, Human Science Press, New York, 1984; pp. 39–58.
22. Hamill PV, Johnson CL, Reed RB, Roche AF, Moore WM. Physical growth: National Center for Health Statistics percentiles. *Am J Clin Nutr* 1986; **32:** 607–629.
23. Ikeda Y, Higurashi M, Hirayama M, Ishikawa N. A longitudinal study on the growth of stature lower limb length in Japanese children with Down's syndrome. *J Ment Defic Res* 1977; **21:** 139–151.
24. Cronk CE, Crocker AC, Pueschel SM, Shea AM, Zackai E, Pickens G, Reed RB. Growth charts for children with Down's syndrome: 1 month to 18 years of age. *Pediatrics* 1988; **81:** 102–110.
25. Baumgartner RN, Roche AF, Himes JH. Incremental growth tables: supplementary to previously published charts. *Am J Clin Nutr* 1986; **43:** 711–722.
26. Rarick GL, Wainer H, Thissen D, Seefeldt VA. A double logistic comparison of growth patterns of normal children and children with Down's syndrome *Ann Hum Biol* 1974; **2:** 339–346.
27. Rundle S. Anthropometry: a ten-year survey of growth and sexual maturation. In *Mental Subnormality: Modern Trends in Research*, edited by Richards, BW, Pitman Medical and Scientific, London, 1970; pp. 68–118.
28. Pueschel SM. Health concerns of persons with Down's syndrome. In *New Perspectives on Down's Syndrome*, edited by Pueschel SM, Paul H. Brookes, Baltimore, 1987, pp. 113–133.
29. Buday J. *Growth and Physique in Down's Syndrome Children and Adults*, Humanbiologia Budapestinensis, Budapest, 1990.
30. Pueschel SM, Orson JM, Boylan JM, Pezzullo JC. Adolescent development in males with Down's syndrome. *Am J Dis Child* 1985; **139:** 236–238.
31. Thelander HE, Pryor D. Abnormal growth patterns and development in mongolism. *Clin Pediatr* 1966; **5:** 493–499.
32. Rarick GL, Rapaport IF, Seefeldt V. Long bone growth in Down's syndrome. *Am J Dis Child* 1966; **112:** 566–571.
33. Greulich WW, Pyle I. *Radiographic Atlas of Skeletal Development of the Hand and Wrist*. Stanford University Press, Stanford, 1959.

34. Roche AF. Skeletal maturation and elongation in Down's disease (mongolism). *Eugene Rev* 1967; **59:** 11–21.
35. Chumlea WC, Cronk CE. Overweight among children with trisomy 21. *J Ment Defic Res* 1982; **25:** 275–279.
36. Cronk CE, Chumlea WC, Roche AF. Assessment of overweight in children with trisomy 21. *Am J Ment Defic* 1985; **89:** 433–436.
37. Roche AF. Difficulties associated with studies of growth as a developmental indicator in mentally retarded children. *Proc First Internat Assoc Sci Study Ment Defic* 1961; 845–852.
38. Cronk CE, Reed RB. Canalization of growth in Down's syndrome. *Hum Biol* 1981; **53:** 383–398.
39. Waddington CH. *The Strategy of Genes*, George Allen & Unwin, London, 1957.
40. Lampl M, Emde RN. Episodic growth in infancy: a preliminary report on length, head circumference, and behavior. In *Levels and Transitions in Children's Development: New Directions for Child Development*, edited by Fischer KW, Jossey-Bass, New York, 1983; pp. 21–36.
41. Hermanussen M, Geiger-Benoit K, Burmeister J, Sippell WG. Periodical changes of short term growth velocity ('mini growth spurts') in human growth. *Ann Hum Biol* 1988; **15:** 103–110.
42. Cicchetti D, Sroufe LA. The relationship between affective and cognitive development in Down's syndrome infants *Child Dev* 1976; **47:** 920–926.
43. Tanner JM, Healy MJR, Lockhart RD, MacKenzie JD, Whitehouse RH. Aberdeen Growth Study I: the prediction of adult body measurements from measurements taken each year from birth to 5 years. *Arch Dis Childh* 1956; **31:** 372–381.
44. Cameron N. The methods of auxological anthropometry. In *Human Growth Vol. 3: Methodology, Ecological, Genetic and Nutritional Effects on Growth*, edited by Falkner F, Tanner JM, Plenum Press, New York, 1986, pp. 1–46.

Discussion of Presentation

CHRISTINE CRONK
(Edited by D.A. Snider and Salvador Castells)

Z. Laron: I have three questions: What was the skeletal maturation in the children with Down's syndrome in the various groups studied? Can you differentiate between children who grew in institutions versus those children who grew in love-caring homes? Because if we feel the psychological factor is important in growth, then certainly we ought to be able to distinguish that, too. Also, what was the sexual development of those who had a flattening in their growth?

C. Cronk: In the Children's Hospital sample, which provided the best data, we had delays at two years of about seven months, at three years of about eight months skeletal maturation, with greater variability—some being retarded, a few advanced, and some normal. This may be due to a kind of destabilization of the different ossification rates in the different centers where bone age was determined. In general, there is a delay in bone age. I did not analyze those data specifically because bone age was not consistently available for all of the children in the Children's Hospital past the age of three. Children growing in institutions are smaller. This is true not just of children with Down's, but other institutionalized children as well. This is why I relied more on the Children's Hospital sample because they were home-reared and more representative of children with Down's syndrome, today. Because there was limited follow-up, I do not have the data on the secondary sex characteristics of these children. Clearly if there has been a delayed adolescence, the individuals would have been low relative to normal. This may be why the girl who was 2 SD below the mean fell to 4 SD at the time she was seen again.

G. Merriam: Several of the papers today will be talking about growth hormone deficiency in Down's syndrome, and yet if you look at growth data for many children with idiopathic growth hormone deficiency their growth is relatively normal, until the age of one and a half or two years of age. Yet your data show a decrement in the growth of Down's children, even at birth and in the first 12 months of life, suggesting that there has to be more to this than just growth hormone deficiency. If you plot your data on the basis of either growth or growth velocity z scores you can see further change occurring around one and a half to two and a half years of life that show the additional burden of growth hormone deficiency kicking in and causing an additional deficiency at that stage. You talked about the adolescent period, but I'm wondering about the transitional period where we might first expect to see the specific effects of growth hormone deficiency.

C. Cronk: It is pretty much agreed that there are multiple causes for the growth retardation in Down's syndrome. It is not limited to growth hormone deficiency. Addressing your second question, that growth velocities are most markedly delayed between 6 and 24 months of age—it would be an interesting analysis to do, and perhaps someone else has done it, to see if there is a kind of break point, where there is enhanced growth deficiency. What we would have to do is contrast prior to 18 months and post 18 months, and see if this is a significant break point. At this time, I don't see one.

P. Hunt: Have you studied the children who are institutionalized? Because fewer children are being admitted to institutions, but those that are, are severely involved; they are severely and profoundly retarded, usually with serious medical problems as well.

C. Cronk: The data from some of these slides came from out of the past. The Wisconsin data are from the 1960s and 1970s, when institutionalization was the general practice. The old data represent the general population of children with Down's syndrome. If we were to do this today, we would be seeing more severely involved children in institutions. All of my measurements have been with home-reared children, not those in institutions.

3 Growth Characteristics in Prepubertal Children with Down's Syndrome in Cantabria, Spain

C. LUZURIAGA, L.V. GARCIA AND J. FLÓREZ*
Departments of Pediatrics and Pharmacology, University Hospital 'Valdecilla' and Faculty of Medicine, University of Cantabria, Santander, Spain

INTRODUCTION

Children with Down's syndrome (DS) born in Cantabria—a northern region in Spain—are commonly included in a health program designed to prevent and/or detect medical anomalies associated with DS, because good physical conditions are needed to optimize learning and intellectual skills. It is known that children with DS often present a pathological growth with a final short stature (1). The present study was undertaken to evaluate growth in our population, to analyze some factors contributing to a pathological height in adulthood, and to study the degree of bone maturation and nutritional conditions.

METHODS

A population of 50 children with DS (26 males and 31 females) was studied. The mean chronological age (CA) was 8.4 ± 2.86 years (SD) (range: 1.58–14 years). Heights and growth rates were expressed as *z* scores (SDS). SDS were calculated on the basis of CA for height, and of the CA and bone age (BA) for growth rates. The growth rate was calculated over a follow-up period of one year. BA was assessed in 44 children by the same investigator, following the method of Greulich and Pyle (2). The differences between BA and CA were consistently evaluated. The predicted adult height was calculated by the method of Bayley-Pinneau (3) in girls ($n = 18$) and boys ($n = 16$), and the predicted height was compared to the target height.

The nutritional condition and its relation to the diet was assessed by determining the weight for height related to each age and sex, and expressed as a

*Address for correspondence: Department of Pharmacology, Faculty of Medicine, University of Cantabria, Avda. Herrera Oria s/n, 39011 Santander, Spain.

Growth Hormone Treatment in Down's Syndrome. Edited by S. Castells and K.E. Wisniewski

percentage according to the chronological standards of Tanner. Percentages between 80% and 120% were considered as normal, and above 120% as obese (4).

RESULTS

Since the auxiological data in both populations—males and females—did not show statistical differences (Table 1), all data were pooled in a single group. However, common anomalies associated with DS were also considered, such as congenital heart disease (CHD) (moderate in six and severe in six), thyroid disease (hyperthyroidism in one, hypothyroidism in six), and diabetes mellitus associated with gluten intolerance in one.

Height was abnormal (height SDS < -2.0) in 23 children (41.1%) including six patients with severe CHD, four with hypothyroidism, and one with diabetes mellitus. Borderline height deviation (SDS from -1.7 to 2.0) was present in seven, including one with hypothyroidism, and six with no additional anomaly. Five children with moderate CHD had a height SDS close to -1.0, and four who did not present any anomaly showed SDS = 0.

The mean value of the growth rate, studied in 35 children, was 5.1 ± 1.6 cm per year (range 1.4–9.4). In relation to the CA, the growth rate appeared abnormal (lower than the 3rd percentile) in 11 patients (26.8%), including some children with severe CHD, four with hypothyroidism and others with no associated anomalies. Growth rate assessed in SDS for BA was also below the third percentile in nine children (24.2%).

In order to evaluate the bone maturation, the population ($n = 44$) was divided into three groups according to the CA: I ($n = 5$, age less than six years), II ($n = 14$, age six to eight years), and III ($n = 25$, age more than eight years). As shown in Table 2, bone maturation appeared to increase faster compared to CA.

In 16 boys, the mean predicted final height was 160.0 ± 5.7 cm, and the target height was 169.5 ± 3.7 ($p < 0.01$). SDS were -1.73 ± 0.6 and -0.56 ± 0.65 for predicted and target heights, respectively ($p < 0.001$). In 18 girls, the mean predicted final height was 149.5 ± 4.4 cm, and the target height

Table 1. Growth characteristics

	CA height SDS	CA height velocity SDS	BA height velocity SDS	Weight/height (%)
Males	−1.71 ± 0.84	0.01 ± 2.64	0.0 ± 1.8	110 ± 12.4
Females	−1.70 ± 1.22	−0.62 ± 1.36	−0.9 ± 1.2	112 ± 15.6
Significance	n.s.	n.s.	n.s.	n.s.

was 159.0 ± 4.2 cm ($p < 0.001$). SDS were −1.18 ± 1.02 and −0.49 ± 0.69 for predicted and target heights, respectively ($p < 0.01$). Individual analysis showed that 10 boys (62.5%) and 10 girls (55.5%) would attain a lower final height than the third percentile (Tunner tables).

Overweight (weight for height percentage higher than 120%) was observed in 12 children (21.05%), two of whom were hypothyroid. Undernourishment was never detected.

DISCUSSION

Our results confirm previous studies (1) showing the prevalence of retarded growth rate and low height in individuals with DS, starting at an early age. The influence of additional anomalies associated with DS, such as CHD and hypothyroidism, are also emphasized (1, 5). However, only few studies have considered the impact of bone maturation on height. In children under 36 months, a maturation delay is frequently observed (6). On the other hand, an increase has been reported in adolescents and young adults, which has been related to the rise in gonadotropins and testosterone during early pubertal development (7). The 'adrenarche' phase of pubertal maturation is another factor to be considered for changes in bone maturation (8). For the purpose of studying bone maturation, and on the basis of previous studies showing that the 'adrenarche' appears at about eight years (9), our population was divided into groups above and below this age. For children below six years, BA was consistently lower than CA (see Table 2). But as CA increased, the differences between BA and CA were positive in 64% and 80% of the groups six to eight and above eight years, respectively. It is therefore possible that additional factors related to adrenal secretion may accelerate bone maturation in DS, thereby contributing to a shorter height of the individual.

Obesity was found only in 21% of the children. Most of the children included in this study were closely controlled and the families received nutritional assessment.

Table 2. Comparisons between CA and BA

Groups	CA (years)	BA (years)	BA − CA	BA − CA > 0	BA − CA > 1
I ($n = 5$) <6 years	3.4 ± 1.1	2.98 ± 0.9	−0.4 ± 0.4	0	0
II ($n = 14$) 6–8 years	7.1 ± 0.5	6.95 ± 0.89	0.1 ± 0.7	8	1
III ($n = 25$) >8 years	10.2 ± 1.6	10.4 ± 1.7	0.2 ± 0.7	16	4

CONCLUSION

Short stature and low growth velocity were observed in a high percentage of children with DS, regardless of the presence of associated anomalies such as CHD and hypothyroidism. The predicted height is commonly lower than target height in both males and females. An acceleration of BA may appear after the age of six to eight years, thus contributing to a shorter final height. Factors contributing to this acceleration should be investigated.

REFERENCES

1. Cronk C, Crocker AC, Pueschel SM, Shea AM, Zackai E, Pickens G, Reed RB. Growth charts for children with Down's syndrome: 1 month to 18 years of age. *Pediatrics* 1988; **81:** 102–110.
2. Greulich WW, Pyle SI. *Radiographic Atlas of Skeletal Development of the Hand and Wrist*, 2nd edn, Stanford University Press, Stanford, 1959.
3. Bayley N, Pinneau SR. Tables for predicting adult height from skeletal age: revised for use with the Greulich–Pyle hand standards. *J Pediatr* 1952; **40:** 423–441.
4. Cole JJ. A method for assessing age-standardized weight–height in children seen cross-sectionally. *Ann Hum Biol* 1979; **6:** 249–268.
5. Hestnes A, Stovner LJ, Husoy O, Folling I, Fougner KJ, Sjaastad O. Hormonal and biochemical disturbances in Down's syndrome. *J Ment Defic Res* 1991; **35:** 179–193.
6. Cronk C, Anneren AG. Growth *Biomedical Concerns in Persons with Down's Syndrome*, edited by Pueschel SM, Pueschel JK, Paul H. Brookes, Baltimore, 1992; pp. 19–37.
7. Pueschel SM, Orson JM, Boylan JM, Pezullo JC. Adolescent development in males with Down's syndrome. *Am J Dis Child* 1985; **139:** 236–238.
8. Sizonenko PC. Endocrinology in preadolescents and adolescents. *Am J Dis Child* 1978; **132:** 704–712.
9. Ferrández A. Maduración endocrina. *An Esp Pediatr* 1983 (Suppl) **18:** 101–108.

4 Microcephaly and Neurological and Neuropathological Abnormalities in Children with Down's Syndrome

K.E. WISNIEWSKI,[1,2*] I. BEAULIEU,[1,2] M. BOBINSKI,[1,3] M.H. LEE[1] AND S. CASTELLS[1,2]

[1]*New York State Institute for Basic Research in Developmental Disabilities, Staten Island, New York Department of Pathological Neurobiology, and* [2]*State University of New York Health Science Center at Brooklyn, New York, USA;* [3]*Department of Anatomy, School of Medicine, Gdansk, Poland*

INTRODUCTION

All individuals with Down's syndrome (DS) have neurological abnormalities, e.g. different degrees of developmental delay and of deficit in language and intellectual function, unusual personality, and occasional seizures. The most characteristic neurological feature of children with DS is mild to moderate mental retardation and impairment of fine and gross motor coordination. In addition, a universal feature of children with DS is central hypotonia, which manifests as decreased reflexes (such as Moro and parachute) during early infancy but tends to improve with age. Other neurological findings may be present, e.g. ocular nystagmus, conductive and/or sensorineural hearing impairment, and complications associated with atlantoaxial subluxation, predisposing to upper cervical spinal cord injury.

Also, it has been found previously that individuals with DS are of short stature and have microcephaly (1–4), both of which might occur secondarily to human growth hormone (hGH) deficiency (5, 6), neuropathological abnormalities (3, 7–9), and chromosomal defect (trisomy of chromosome 21, causing abnormal gene dosage effect) (10).

The neuropathological abnormalities that occur in individuals with DS are prenatal retardation of neurogenesis, or migration defect, mainly of the neurons of the granular cell layers (II and IV), with a 20–50% decrease in the

*Address for correspondence: Department of Pathological Neurobiology, New York State Institute for Basic Research in Developmental Disabilities, 1050 Forest Hill Road, Staten Island, NY 10314, USA.

Growth Hormone Treatment in Down's Syndrome. Edited by S. Castells and K.E. Wisniewski

number of neurons throughout the neocortex, and defects of prenatal and postnatal synaptogenesis (3, 7, 8) and hypothalamic abnormalities (11).

Our hypothesis, elaborated upon herein, is that neurological and neuropathological abnormalities occur secondarily to hGH abnormalities and gene dose effect. When children with DS were treated with hGH, they showed increased head and brain size and somatic growth (9, 6, 12).

To prove the hypothesis that hGH may promote increased neuronal and synaptic proliferation and brain growth by nerve growth factors (NGFs) and any other growth factors in the brain, experimental models (e.g. tadpoles, rats (13–15), snail dwarf mouse, rats (16–19), and trisomy 16 mouse (Ts16)) were studied (20). Also, the general rules and regulations of the different stages of development and maturation of the central nervous system (CNS) in children with DS will be discussed.

GENERAL RULES AND REGULATIONS IN DEVELOPMENT AND MATURATION OF THE CENTRAL NERVOUS SYSTEM

Differentiation and maturation of the CNS follow discrete timetables that differ depending on several factors, such as the area of nervous structure considered (e.g. the spinal cord matures before the cerebral cortex, and the latter before the cerebellum) and the animal species utilized for study (e.g. in humans and guinea pigs, most neuronal differentiation occurs prenatally, whereas in others, such as rats and mice, it occurs postnatally). Several developmental events are common to all neural cells and reflect growth processes that occur in all cells; others are processes that are unique to neurons and glial cells. Such developmental timetables (e.g. proliferation of undifferentiated cells followed by their migration and subsequent differentiation at the final destination site) are well established. In addition to general, growth-related processes, e.g. protein, DNA and RNA synthesis, neural cell development includes specific events such as neurite growth, synaptogenesis, development of spontaneous and evoked electrical activity, specification of neurotransmitter function, and myelinogenesis.

Finally, when the CNS is built, about 100 billion signalling neurons are present. They are organized into a network and regions, and some have specialized functions. In the prenatal and postnatal stages of CNS development, the endocrine glands have an important influence, e.g. deficiency or excess of endocrine secretions profoundly alters each of the morphological, functional and biochemical events. GH and/or nerve growth factors and thyroid hormone may serve as signals for synchronization of developmental sequences. Any desynchronization of the signal or signals during critical periods of CNS maturation leads to a series of consequences, which may immediately affect development and maturation and, at later ages, may influence accelerated aging as well, e.g. reduced neurogenesis, synaptogenesis, increased cell death and impaired myelinogenesis.

The brains of individuals with DS have been examined by many researchers from the beginning of this century (21). The studies have continued (22), increasing in number particularly in recent years (7–9, 23–31). Presently, morphological features of the CNS at the prenatal and postnatal stages (at the gross, light microscopic and ultrastructural levels) show abnormalities. In DS, abnormalities are known to be present in brain shape; brain weight; cortical organization, including number of neurons, synaptic density, and their particular features: and myelination of brain pathways. Degenerative changes also occur.

PRENATAL DEVELOPMENT AND MATURATION OF THE CENTRAL NERVOUS SYSTEM IN CHILDREN WITH DOWN'S SYNDROME

Observations concerning prenatal development of the CNS in DS are rare. Sylvester (32) found the brain to be smaller and the hippocampus less mature in one DS case than in a control case at 18 weeks of developmental age. A more exhaustive morphometric study performed by Schmidt-Sidor *et al.* (31) in 15 DS and eight non-DS, age-matched fetuses from 15 to 22 weeks gestational age demonstrated that during this span of time no apparent differences in brain growth and maturation occurred. In both groups, the heads of fetuses were brachycephalic, with similar circumferences and configurations of frontal and temporal lobes and almost the same size cerebellum and brain stem. In both groups, the development of cerebral and cerebellar cortices as well as myelination of spinal cord and brain stem tracts occurred at the same stage. Biochemical studies of the composition of gangliosides and neuronal cell adhesion molecules and the activity of acetylcholinesterase also did not show differences between DS and non-DS fetal brains at this stage of development (33). The first difference between the DS and non-DS groups in brain weight was observed by Wisniewski *et al.* (8) and Schmidt-Sidor *et al.* (31) at the last trimester and at birth. They estimated the brain weights of newborns with DS to be lower than those of non-DS controls, although they could be considered to be in the lower-normal weight range. Benda (25, 34) considered the brain weights of newborn with DS to be normal.

Microscopic examination has permitted the discovery of some differences between the cerebral cortices of newborns with and without DS. Neuronal density in the occipital/temporal and frontal areas was demonstrated to be lower in individuals with DS than in persons without DS at birth (7, 8). The morphology of neurons, their dendrites, and synaptic spines in these areas, estimated by the Golgi-rapid method by Takashima *et al.* (35), appeared until birth to be similar in both groups. Ultrastructural studies in cases of DS revealed decreased synaptic density, presynaptic length, and average surface area per synaptic contact at the first day of life (8). Marin-Padilla (36) and Petit *et al.* (37) also found anomalies in synaptic morphology but only a slight reduction in their density. The disturbed electrical membrane properties of the fetal brain were observed very early, at 16 weeks of gestational age (26).

In summary, the first observed differences in brain between individuals with DS and persons who do not have DS were in newborns, but they may start prenatally during the second half of fetal development. In data related by various investigators, which on some points were conflicting, depending on the population examined and the methods used for examination, the degree of observed changes was moderate. Nevertheless, particular features found in the neuronal population of the cerebral cortex in newborns with DS testify that the changes may be present from the time of conception but are more visible in the later prenatal stages of brain differentiation.

MICROCEPHALY AND POSTNATAL DEVELOPMENT AND MATURATION OF THE CENTRAL NERVOUS SYSTEM IN CHILDREN WITH DOWN'S SYNDROME (FROM BIRTH TO EARLY CHILDHOOD)

After birth, the changes occurring in the developing brains of children with DS become much more evident than changes observed during prenatal life. They are particularly accentuated during early infancy and early childhood.

Early infancy is the time when general neuropsychological development of children with DS is slow. Differences from the normal population increase with age: usually, levels of cognitive function and speech development in children with DS are lower than in controls.

Head circumference (HC) becomes microcephalic, being as much as 3.0 SD lower than the expected value during the first two years of age (38). We have reported that about 50% of children with DS up to three years of age (4) and 80% up to five years of age (3) are microcephalic. Our examination of occipitofrontal circumferences (OFC) of 720 children with DS (ages from birth to 60 months) showed microcranium after mid-infancy in most cases. Twenty per cent of children with DS had an OFC in the lower-normal range. When compared with the mean HC in controls from the Nellhause study (39), the mean HC of children with DS is significantly lower, especially after mid-infancy (Figs 1 and 2; Table 1). Also, HC curves of the children with DS varied in the same age group and may be 2 SD below (Figs 3 and 4) or 2 SD above (Fig 5) the norm. The girls usually had slightly smaller HCs than the boys.

The deficit in brain weight in infants with DS is parallel to the deficit in HC. The differences in HC and brain weight between children with and those without DS have been observed by some investigators to increase from birth. This difference is particularly visible between two and four years of age, and

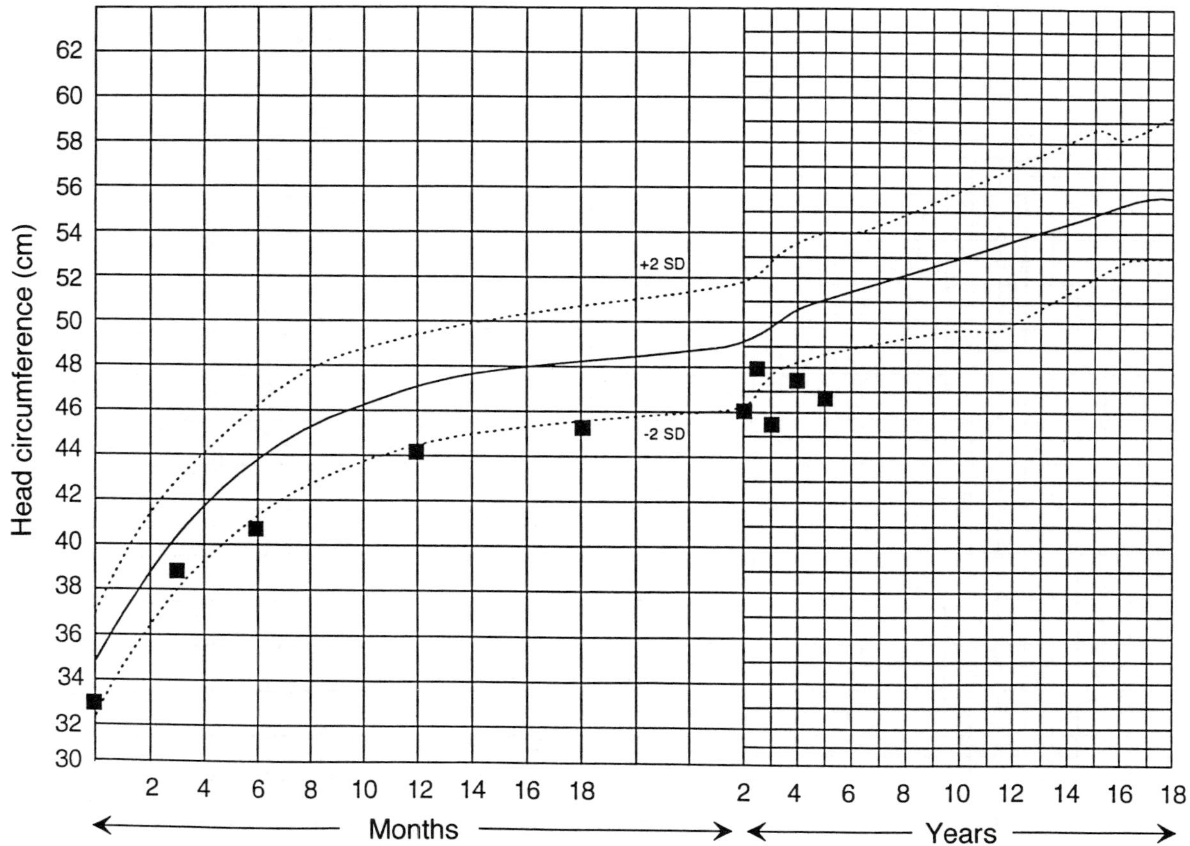

Figure 1. The mean of head circumferences of 390 boys with DS; see dark square (ages birth to 60 months against controls) (39). The number of cases and the mean ± SD are given in Table 1

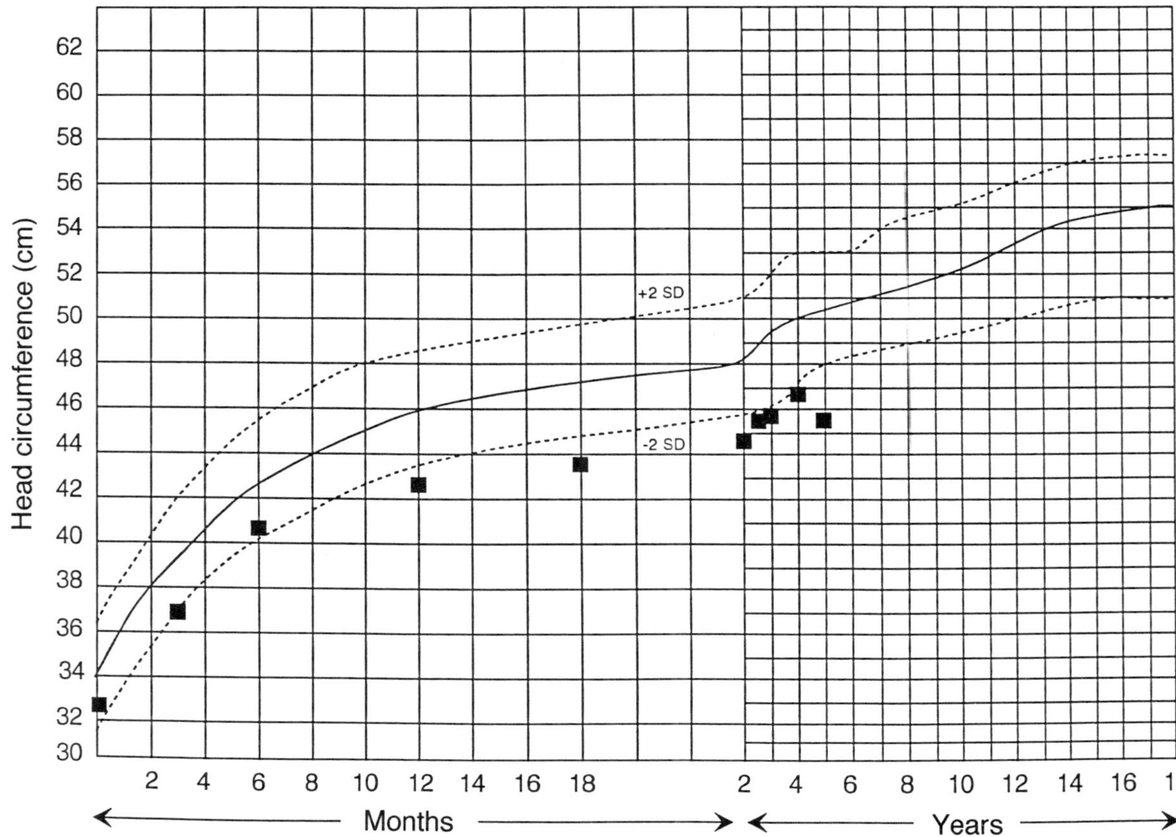

Figure 2. The mean of head circumference of 330 girls with DS; see dark square (ages birth to 60 months against controls) (39). The number of cases and the mean ± SD are given in Table 1

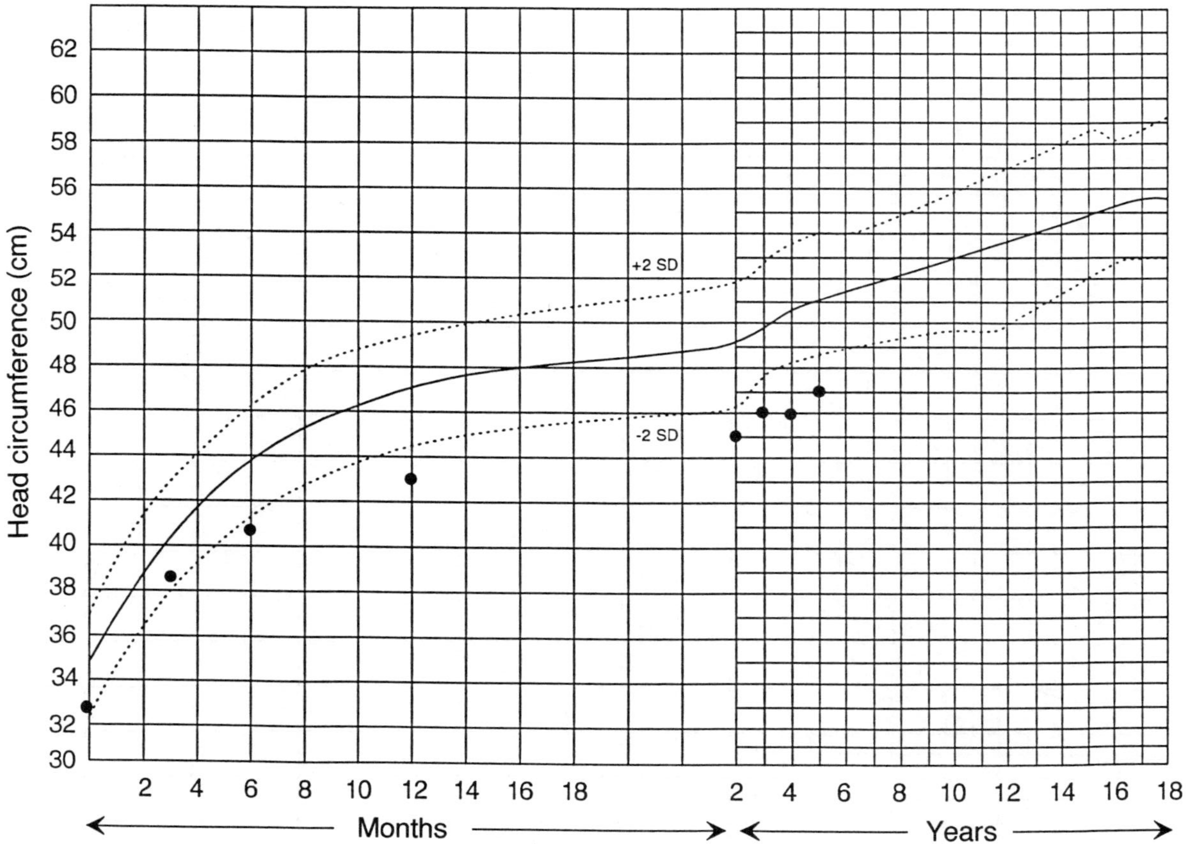

Figure 3. Head circumference of boy with DS was plotted on Nellhause curve. At birth, the HC was in the lower range of normal, but after six months it represented microcephaly

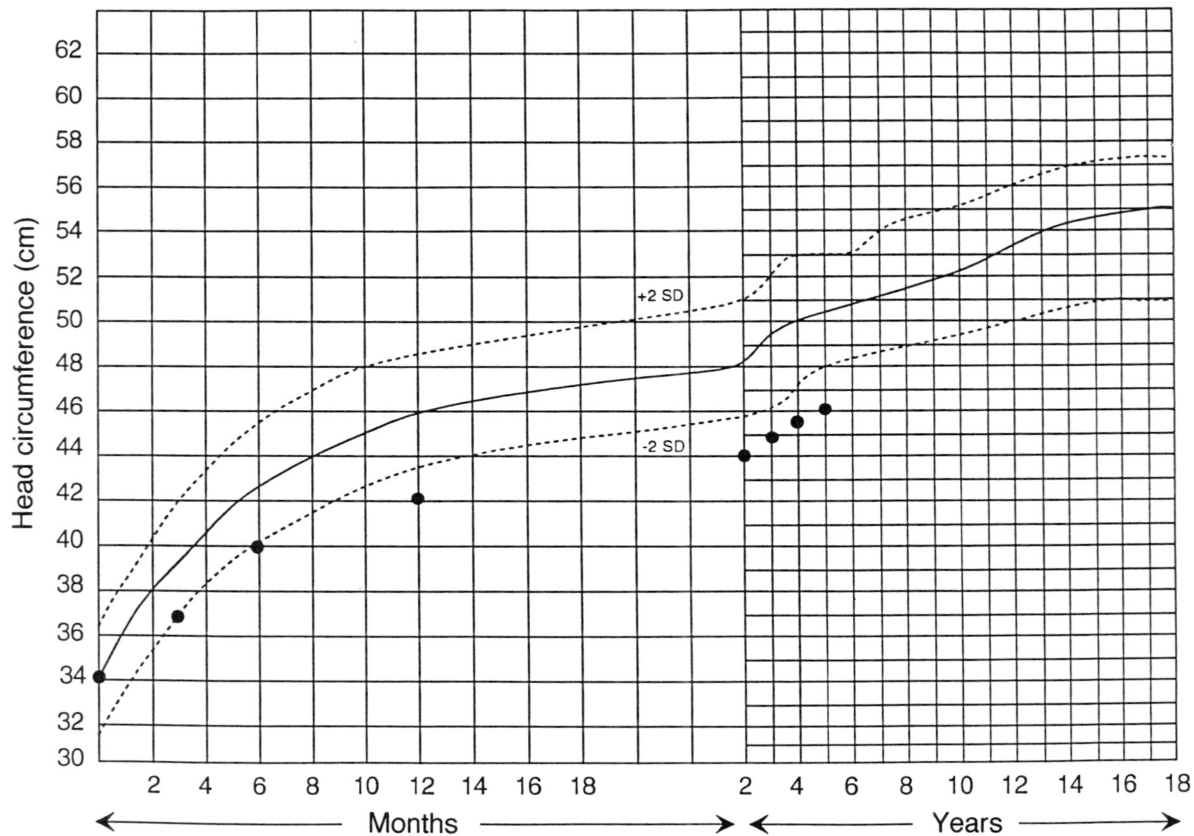

Figure 4. Head circumference of girl with DS; see dark circle, which was plotted on Nellhause curve. At birth, the HC was at the mean, but at six months it was at a lower range and at 12 months represented microcephaly

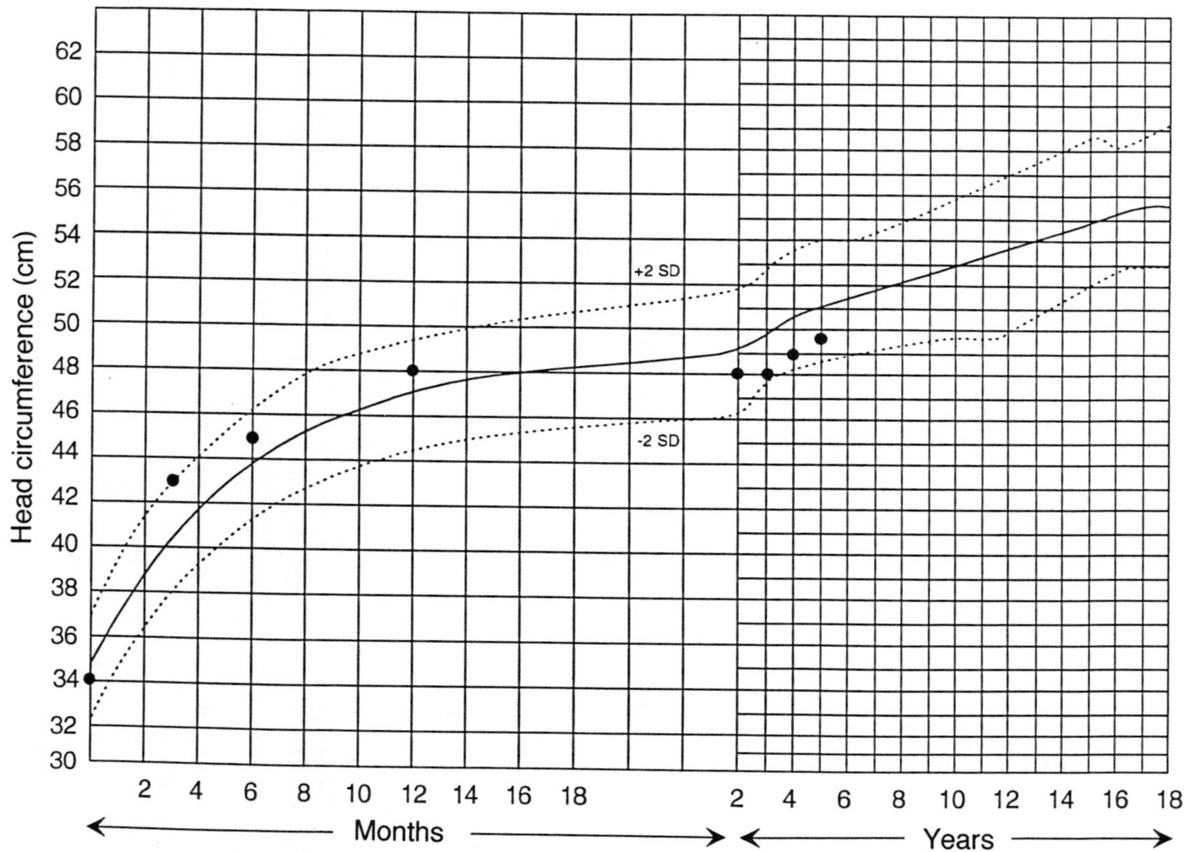

Figure 5. Head circumference (HC) of boy with DS that was plotted on Nellhause curve. At birth, the HC was at 98%, but at 60 months of age it was in the lower normal range

Table 1. Mean and standard deviation (SD) of HC measurements of 720 DS cases. These measurements are plotted on the Nellhause curve (390 boys and 330 girls) (see Figs 1 and 2)

	Boys			Girls		
Age (months)	IBR[a] n	C&P[b] n	Mean ± SD	IBR[a] n	C&P[b] n	Mean ± SD
Birth	7	18	32.7 ± 0.7	3	16	32.7 ± 1.5
3	5	48	38.6 ± 1.4	7	39	37.0 ± 1.3
6	4	46	40.6 ± 1.8	2	38	40.8 ± 2.0
12	5	45	44.9 ± 0.9	9	37	42.6 ± 1.2
18	4	47	45.1 ± 1.6	9	35	43.8 ± 1.2
24	10	44	45.9 ± 0.6	6	34	44.4 ± 1.9
30	0	49	47.8 ± 1.3	3	38	45.3 ± 1.0
36	9	47	45.3 ± 0.3	9	37	45.5 ± 0.5
48	6	–	47.3 ± 1.3	5	–	46.7 ± 2.1
60	4	–	46.5 ± 0.5	3	–	45.5 ± 0.5

[a]New York State Institute for Basic Research in Developmental Disabilities.
[b]Cronk and Pueschel (1).

it has been concluded that the 'most evident brain weight retardation is seen after mid-infancy' (8, 28, 31). We have studied the brain weight of 101 children with DS and 80 children who do not have DS between birth and 60 months of age. The increases in brain weight with age were greater in the children without than with DS. When the brain weights of both groups were compared with previously established norms (40, 41), it was found that the number of children with DS with brain weight below normal increased with age (Figs 6–8). The degree of deficit in brain weight was evaluated as 24% in infants with DS (24) and 30–50% in children with DS (8) of the brain weights of control age-matched subjects. Therefore, the cases of DS can often be diagnosed as microcranial and microcephalic. As a result of examining both DS and non-DS brains, Schmidt-Sidor *et al.* (31) verified that lower brain weight is associated with gastrointestinal malformations and congenital heart disease.

The shape of DS brains observed in the majority of cases presents some particular features. The first one observed by many authors—shortened anteroposterior dimension—was attributed to the reduction of the frontal lobes (8, 28, 31, 34) and was not evident in another study (30). Superior temporal gyri in one or both hemispheres were often found to be narrow. This phenomenon was found in 33% of cases investigated by Schmidt-Sidor (31) and in 50% of brains examined by others (42, 43). The smaller size of the cerebellum and brain stem in individuals with DS than in individuals who do not have DS has often been reported (3, 24, 43, 44). Even the ratio of the weight of the cerebellum and brain stem to the weight of cerebrum was found

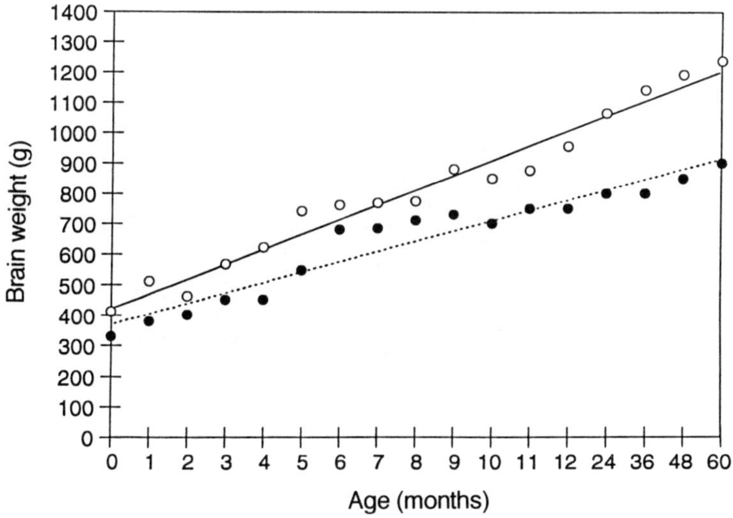

Figure 6. Comparison of brain weights of boys with DS and of control subjects (Baker 1967). Note that weights of DS brains are lower than weights of control subjects. Open circle = control; closed circle = boys with DS

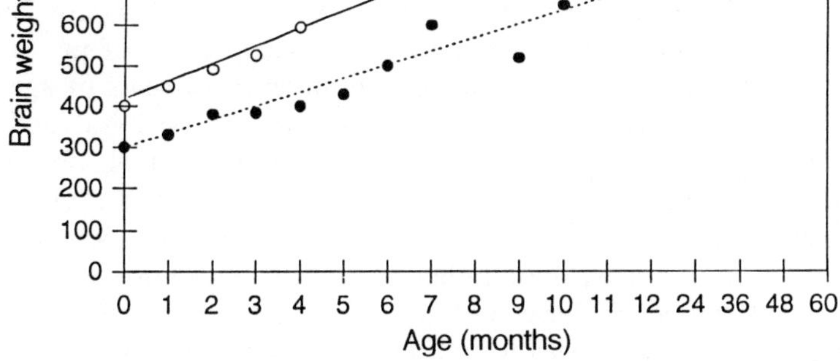

Figure 7. Comparison of brain weights of girls with DS and of control subjects (Baker 1967). Note that weights of DS brain are lower than weights of control subjects. Open circle = control; closed circle = girls with DS

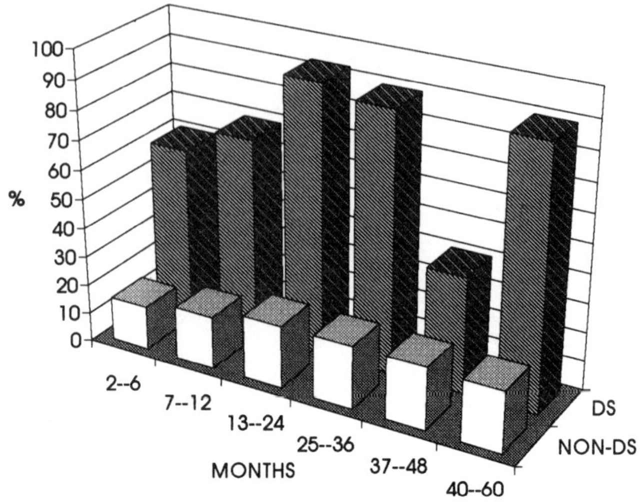

Figure 8. Percentage of brains with weight below normal range. Brain weights from to 60 months of age of 101 individuals with DS compared with those of 80 brains of age-matched controls without DS. The number of subjects differed in each group: two to six months, 25; 7–12 months, 16; 13–24 months, 27; 25–36 months, nine; 37–48 months, five; 40–60 months, nine. Note that the percentage of low brain weight increases with age in comparison to control. The number of cases in the group 37–48 months was relatively low (five cases); therefore, most likely, the difference was not so distinct

by Crome *et al.* (24) to be 1 : 9 in persons with DS and 1 : 7 in control subjects without DS.

The developmental disability that occurs in children with DS prompted studies of neuronal density in various brain structures, particularly in the cortex. Reduced neuronal density in DS was reported in very early studies (21, 45); Sylvester (32) described a decreased total number of neurons in the hippocampus, and Zellweger (42) noted difficulty in recognizing cortical layers because of cytoarchitectonic changes. Morphometric study of neuronal density in three areas of the cerebral cortex from birth to 14 years of age demonstrated that densities were significantly lower in the occipital and frontal cortices and hippocampus of children with DS than in control subjects (7, 8). Examination showed that decreased neuronal density in the visual cortex occurred particularly in layers II and IV. In the frontal cortex, which normally contains few granular cells, the number of neurons was not significantly less than in control subjects. These types of changes are in agreement with the results of Ross *et al.* (27), who found deficits of small neurons in the cortex of the brains of two individuals with DS.

In DS, not only are the cortical neurons reduced in number, but the number of synaptic spines is also reduced. Such observations were made by Takashima *et al.* (35, 46) with use of the Golgi-rapid method; they found the dendritic spines after the fourth postnatal month to be not only reduced in number but also very thin. The results of Becker *et al.* (47), performed on the visual cortex in children with DS from four months to five years of age, were similar. Suetsugu and Mehraein (48) observed the diminished number of synaptic spines in the parahippocampal and cingulate gyri in their study subjects with DS, ranging from 3 to 25 years of age. The changes were found not only to persist to adulthood but also to overlap with the aging process, as was found by Takashima *et al.* (46) in individuals with DS older than 30 years of age.

Ultrastructural observations of synaptic morphology in the visual cortex from the first day of life until 18 years of age were made by Wisniewski *et al.* (8) with the use of morphometric methods. They confirmed that synaptic length and surface area per synaptic contact were less in individuals with DS than in control subjects without DS. Dendritic spines receive the majority of excitatory synaptic input onto neurons, and abnormal spine morphology can significantly alter the normal flow of current through the neuron. In DS, there are few spines, and those that are present tend to be long, with thin necks. Long thin necks have higher resistance to current flow than those that are short and stubby, leading to a greater voltage drop from the spine head to the dendrite shaft. This would distort synaptic input such that less voltage would reach the axon hillock, where voltage changes throughout the entire neuron are summated and give rise to action potentials. Interestingly, in normal prenatal brains, long thin spines are the rule; postnatally, spine morphology becomes more varied. The abnormal persistence of long thin dendritic spines in brains of individuals with DS and other types of mental retardation may lead to abnormal susceptibility to seizure. Brains of individuals with DS also show a higher percentage of 'primitive' synaptic profiles on ultrastructural analysis (37). The ontogenesis of spine formation in DS suggests that before birth normal spines are present, but that after birth there is an arrest of spine development that parallels cognitive dysfunction (35). In addition, the elaboration of the dendritic tree appears relatively normal until about six months after birth, after which there is an apparent atrophy of dendrites (47). The presence of abnormal filopodium-like structures on the soma of cortical neurons of infants with DS has been reported (49); the functional consequences of these structures are unknown, but such processes might support aberrant excitatory neuronal transmission and contribute to a hyperexcitable state. The altered neuronal structure and inhibitory neurotransmission (fewer inhibitory interneurons) could be the cause of seizures in children with DS. The decreased number of neurons and abnormal synapsis could be the cause of cognitive dysfunction.

CNS gray matter abnormalities are diffuse, and they extend to other regions. Our studies of the arcuate (ARC) and ventromedial (VMN) nuclei of the hypothalamus, which are the sites of the highest concentrations of the growth hormone-releasing factor, revealed characteristic changes in this region in individuals with DS. We found that the number of neurons per square

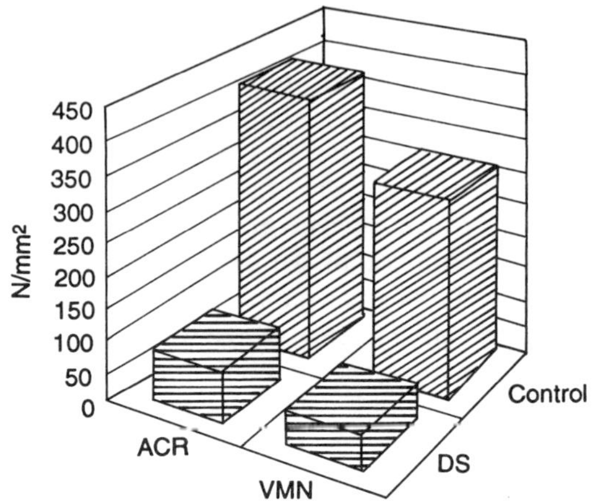

Figure 9. Number of neurons per square millimeter (N/mm^2) in the ARC and VMN in control and DS groups. The diagram shows a significant decrease in the neuronal density in both nuclei of the individuals with DS

millimeter in the ARC and VMN was extremely diminished in brains of persons with DS (Fig 9). These results could suggest a decline in the secretory activity of the ARC and VMN (11).

Myelination of the cerebral pathways—another important parameter of brain maturation—was examined by Wisniewski and Schmidt-Sidor (9). They found myelination delay in 22% of examined cases. The myelination was most evidently retarded during the first year of life—the time during which myelination within the cerebral hemispheres usually occurs most intensively—but in some cases it was seen even later. The topography of tracts with retarded myelination consisted of 'tracts with late beginning and protracted cycle of myelination' (9). The occurrence of impaired gross and fine motor coordination in children with DS most likely is secondary to the protracted cycle of myelination. This phenomenon was observed in corticopontine and corticospinal tracts, corona radiata, corpus callosum, and frontal and temporal lobes. Less advanced myelination was also found in the U-fibers in 50% of research subjects.

INFLUENCE OF TREATMENT WITH GROWTH HORMONE, PAST AND PRESENT

Zamenhof (13–15) has investigated the possibilities that brain function increases with increased numbers of neurons by using pituitary extract (acidophile cells of the anterior lobe of the hypophysis). He was able to stimulate the proliferation of cells in the early period of brain development and maturation in tadpoles and rats. Pregnant animals were subjected to intraperitoneal injections of GH. The numbers of cells in the vicinity of the foramen of Monro were 44–126% higher in animals treated than in controls (Table 2). Other studies, in which different animals were treated prenatally with GH, showed similar findings. In Noguchi's (16–19) investigations of the motor cortex (area 6 of Caviness) in Snell dwarf mouse, which were carried out by means of a modified Golgi silver impregnation method, the pyramidal neurons located in layer V were found to have small perikarya and short primary dendrites with sparse branchings. After administration of bovine growth hormone (bGH) and thyroxine (T_4) during the first 20 days of post-natal life, this retarded neuronal growth was restored to normal: neuronal perikarya were enlarged, and the dendrites were thicker, with denser branchings. These results were confirmed in the sections stained for neuron-specific enolase (NSE). His findings demonstrated that exogenous GH and T_4 will (a) enhance neuronal growth in the Snell dwarf cerebrum and (b) restore neuronal arborization to normal.

The Ts16 mouse has for the last 20 years been known to be an animal model of DS (50–52). Because the Ts16 mouse has various severe malformations, it lives only up to the 17th to 19th day of fetal life. Despite this brevity, the Ts16 mouse has been found to have similar cortical dysgenesis abnormalities in the brain (53) as in the brain of children who have DS (3). Our experimental work on Ts16 mice comprised treating them prenatally with hGH to determine if the hypothalamic abnormalities could be diminished or abolished. The arcuate nucleus and the ventromedial nucleus of the hypothalamus, which play a key role in the regulation of GH secretion, were studied (20). The volume of both types of hypothalamic nuclei increased significantly, but the number of cells per square millimeter decreased. Our preliminary results indicate that prenatal treatment of Ts16 mice with hGH increases their somatic and brain weights. Behavioural and neuropsychological testing are needed to better define the correlation between brain volume and cognitive function level.

Table 2. Prenatal treatment with GH causes increased brain size and neuronal proliferation

Authors	Year	Animal	GH
Zamenhof	1941	Tadpoles	Pituitary extract
Zamenhof *et al.*	1966	Rats	Purified GH
Noguchi *et al.*	1984	Snell dwarf mouse	Purified GH
Wisniewski *et al.*	1992	Mouse (C67, Ts16)	Recombinant hGH

REFERENCES

1. Cronk E, Pueschel S. Autopometric studies. In *A Study of the Young Child with Down's Syndrome*, edited by Pueschel SSM, Human Science Press, New York, 1983; pp. 105–141.
2. Wisniewski KE. In Down's syndrome (DS) children, post-natal retardation of head and brain growth (birth-five years). In *Disorders of the Developing Nervous System: Changing Views on their Origins, Diagnoses, and Treatments*, Proceedings of the 1987 Albany Birth Defects Symposium XVIII, edited by Swann JW, Messer A, Liss, New York, 1988; p. 262 (Abstract).
3. Wisniewski KE. Down's syndrome children often have brain with maturation delay: retardation of growth and cortical dysgenesis. *Am J Med Genet Suppl* 1990; **7:** 274–281.
4. Palmer CGS, Cronk C, Pueschel SM, Wisniewski KE, Laxova R, Crocker AC, Pauli RM. Head circumference of children with Down's syndrome (0–36 months). *Am J Med Genet* 1992; **42:** 61–67.
5. Sara VR, Lazarus L. Prenatal action of growth hormone on brain and behaviour. *Nature* 1974; **250:** 257–258.
6. Torrado C, Bastian W, Wisniewski KE, Castells S. Treatment of children with Down's syndrome and growth retardation with recombinant human growth hormone. *Pediatrics* 1991; **119:** 478–483.
7. Wisniewski KE, Laure-Kamionowska M, Wisniewski HM. Evidence of arrest of neurogenesis and synaptogenesis in brains of patients with Down's syndrome (letter). *N Engl J Med* 1984; **311:** 1187–1188.
8. Wisniewski KE, Laure-Kamionowska M, Connell F, Wen GY. Neuronal density and synaptogenesis in the postnatal stage of brain maturation in Down's syndrome. In *The Neurobiology of Down's Syndrome*, edited by Epstein CJ, Raven Press, New York, 1986; pp. 29–44.
9. Wisniewski KE, Schmidt-Sidor B. Postnatal delay of myelin formation in brains from same Down's syndrome infant and children. *Clin Neuropathol* 1989; **8:** 55–62.
10. Epstein J. The consequences of altered gene dosage in trisomy 21. In *New Perspectives on Down's syndrome*, edited by Pueschel SM, Tingey C, Rynders JE, Crocker AC, Crutcher DM, Brookes, Baltimore, 1987; pp. 69–80.
11. Wisniewski KE, Bobinski M. Hypothalamic abnormalities in Down's syndrome. In *The Morphogenesis of Down's Syndrome*, edited by Epstein CJ, Wiley–Liss, New York, 1991; 153–167.
12. Castells S, Beaulieu I, Wisniewski KE, Chapter 12 of present volume.
13. Zamenhof S. Stimulation of the proliferation of neurons by the growth hormone. I. Experiments on tadpoles. *Growth* 1941; **5:** 123–139.
14. Zamenhof S. Stimulation of cortical-cell proliferation by the growth hormone. III. Experiments on Albion rats. *Physiol Zool* 1942: **15:** 281–292.
15. Zamenhof S, Mosley J, Schuller E. Stimulation of the proliferation of cortical neurons by prenatal treatment with growth hormone. *Science* 1966; **152:** 1396–1397.
16. Noguchi T, Sugisaki T, Takamatsu K, Tsukada Y. Factors contributing to the poor myelination in the brain of the Snell dwarf mouse. *J Neurochem* 1982; **39:** 1693–1699.
17. Noguchi T, Sugisaki T, Tsukada Y. Stimulation of Snell dwarf mouse neuronal growth by GH and T_4. *Neurochem Pathol* 1984; **2:** 123–138.
18. Noguchi T, Sugisaki T, Tsukada Y. Microcephalic cerebrum with hypomyelination in the growth hormone-deficient mouse (lit). *Neurochem Res* 1985; **10:** 1097–1106.

19. Noguchi T, Sugisaki T. Abnormal neuronal growth in the little (lit) cerebrum. *Exp Neurol* 1985; **89:** 274–278.
20. Wisniewski KE, Castells S, Wierzba-Bobrowicz T, Bobinski M, Beaulieu I, Lee M, Jenkins JC. Studies of the hypothalamic nuclei of 19 day old mice with trisomy 16, whose mothers were treated prenatally with recombinant human growth hormone. *Acta Neuropathol* 1993 (in preparation).
21. Apert E. Mongolism. *Le Monde Medicale* 1914; **24:** 201.
22. Gans A. Anatomische beobachtungen bei der mongoloiden idiotic. *Nederl Tijdschr Geneesk* 1925; **69:** 922–925.
23. Jacob H. Mongolismus. In *Hanbuch der speziellen pathologischen anatomie und histologie*, edited by Lubarsch O, Henke F, Rossle R, Bd. 13: Nervensystem, edited by Scholz W, teil 4: Erkrankungen des zentralen Nervensystems. Springer-Verlag, Berlin, 1956; pp. 82–98.
24. Crome L, Covie V, Slatter E. A statistical note on cerebellar and brain stem weight in mongolism. *J Ment Defic Res* 1966; **10:** 69–72.
25. Benda CE. *Down's Anomaly* 2nd edn, Grune & Stratton, New York, 1969.
26. Scott BS, Becker LE, Petit TL. Neurobiology of Down's syndrome. *Prog Neurobiol* 1983; **21:** 199–237.
27. Ross MN, Galaburda AM, Kemper TL. Down's syndrome: is there a decreased population of neurons? *Neurology* 1984; **34:** 909–916.
28. Wisniewski KE, Torrado C, Castells S. Treatment in growth hormone deficient Down's syndrome children with recombinant human growth hormone. *Down's Syndrome Papers and Abstracts for Professionals* 1990; **12:** 1–2.
29. Kemper TL. Neuropathology of Down's syndrome. In *The Psychobiology of Down's Syndrome*, edited by Nadel L, MIT Press, Cambridge, MA, 1988; pp. 269–289.
30. Courchesne E. Physioanatomical considerations in Down's syndrome. In *The Psychobiology of Down's syndrome*, edited by Nadel L, MIT Press, Cambridge, MA, 1988; pp. 291–313.
31. Schmidt-Sidor B, Wisniewski KE, Shepard TH, Sersen EA. Brain growth in Down's syndrome subjects: ages 15 to 22 weeks of gestational age and birth to 60 months. *Clin. Neuropathol* 1990; **9:** 181–190.
32. Sylvester PE. The hippocampus in Down's syndrome. *J Ment Defic Res* 1983; **27:** 227–236.
33. Brooksbank BWL, Walker D, Balazs R, Jorgensen OS. Neuronal maturation in the fetal brain in Down's syndrome. *Early Hum Dev* 1989; **18:** 237–246.
34. Benda CE. Mongolism. In *Pathology of the Nervous system*, edited by Minckler J, McGraw-Hill, New York, 1971; **2:** 1361–1371.
35. Takashima S, Becker LE, Armstrong DL, Chan F. Abnormal neuronal development in the visual cortex of the human fetus and infant with Down's syndrome: a quantitative and qualitative Golgi study. *Brain Res* 1981; **225:** 1–21.
36. Marin-Padilla M. Pyramidal cell abnormalities in the motor cortex in a child with Down's syndrome: a Golgi study. *J Comp Neurol* 1976; **167:** 63–81.
37. Petit TL, LeBoutillier JC, Alfano DP, Becker LE. Synaptic development in the human fetus: a morphometric analysis of normal and Down's syndrome neocortex. *Exp Neurol* 1984; **83:** 13–23.
38. Roche AF. The cranium in mongolism. *Acta Neurol Scand* 1966; **42:** 62–78.
39. Nellhause G. Head circumference from birth to eighteen years: practical composite international and interracial graphs. *Pediatrics* 1967; **41:** 106–114.
40. Baker RD. *Postmortem Examination*, Saunders, Philadelphia, 1967.
41. Potter EL, Craig JM. *Pathology of the Fetus and the Infant*, 3rd edn, Year Book Medical Publishers, Chicago, 1975.

42. Zellweger H. Down syndrome. In *Handbook of Clinical Neurology*, Vol. 31, Pt II, North-Holland, Amsterdam, 1977: pp. 367–469.
43. Friede RL. *Developmental Neuropathology*, Springer-Verlag, Berlin, 1989.
44. Gulotta F, Rehder H. Chromosomal anomalies and central nervous system. *Birth Pathol* 1974; **152:** 74–80.
45. Davidoff ML. The brain in mongolian idiocy. *Arch Neurol Psychiatr* 1928; **20:** 1229–1257.
46. Takashima S, Jeshima A, Nakamura H, Becker LE. Dendrites, dementia and Down's syndrome. *Brain Dev* 1989; **11:** 131–133.
47. Becker LE, Armstrong DL, Chan F. Dendritic atrophy in children with Down's syndrome. *Ann Neurol* 1986; **20:** 520–526.
48. Suetsugu M, Mehraein P. Spine distribution along the apical dendrites of the pyramidal neurons in Down's syndrome. *Acta Neuropathol* 1980; **50:** 207–210.
49. Fabregues I, Ferrer I. Abnormal perisomatic structures in non-pyramidal neurons in the cerebral cortex in Down's syndrome. *Neuropathol Appl Neurobiol*, 1983; **9:** 165–170.
50. Oster-Granite ML, Baker C, Ozand PT. Neuroanatomic, ocular, and audio-vestibular malformations in the trisomy 16 mouse. *Pediatr Res* 1983; **17:** 300A.
51. Oster-Granite ML, Gearhart JD, Hatzidimitriou G. Development of the hippocampal formation in the trisomy 16 mouse. *Pediatr Res* 1985; **19:** 328a.
52. Oster-Granite ML, Gearhart JD, Reeves RH. Neurobiologic consequences of trisomy 16 in mice. In *The Neurobiology of Down's syndrome*, edited by Epstein CJ, Raven Press, New York, 1986; p. 137.
53. Sweeney JE, Höhmann CF, Oster-Granite ML, Coyle JT. Neurogenesis of the basal forebrain in euploid and trisomy 16 mice: an animal model for developmental disorders in Down's syndrome. *Neuroscience*, 1989; **31:** 413–425.

5 Diagnosis of Hypothalamic Dysfunction in Down's Syndrome Producing Growth Hormone Deficiency

CARMEN TORRADO[1,2] **AND SALVADOR CASTELLS**[1,3]

[1]*Department of Pediatrics, State University of New York Health Sciences Center at Brooklyn,* [2]*Staten Island University, and* [3]*Institute for Basic Research in Developmental Disabilities, New York, USA*

INTRODUCTION

Children with Down's syndrome (DS) are known to have retarded linear growth along with abnormalities in different organ systems, including nervous and skeletal systems as previously noted (1). The authors have studied the hypothalamic–pituitary function in children with DS with different degrees of developmental disabilities and growth retardation, but without congenital heart disease (CHD). The results suggest that DS children have a growth hormone (GH) deficiency secondary to abnormalities in the stimulatory effect of hypothalamic centers in the secretion of pituitary GH (2).

The aims of this chapter are to review several methods currently available to diagnose GH deficiency and to assess the hypothalamic–pituitary response in children with DS in order to establish the diagnosis of GH deficiency accurately and identify which individuals would probably benefit from GH treatment.

METHODS OF TESTING THE HYPOTHALAMUS

The stimulatory tests can be classified as physiological and biochemical. The physiological tests most often employed are physical exercise and 12 h and 24 h integrated GH concentrations (IC-GH). The exercise test using a bicycle ergometer is cumbersome to do with children, and it cannot be applied with those below the age of five years or in older children with mental retardation. Thus it is not suitable for use with children with DS. There are also reports that exercise stimulatory testing is not reliable, with the percentage of false positives in some studies ranging from 9% to 47% (3–5).

Growth Hormone Treatment in Down's Syndrome. Edited by S. Castells and K.E. Wisniewski

Initial studies of the secretion of GH over 24 h were performed in the 1960s and 1970s mainly for their physiological rather than diagnostic value. Twelve-hour or 24 h IC-GH helps in the diagnosis of GH deficiency when there is variability of GH responses to two different pharmacological tests (6, 7).

The measurement of spontaneous GH secretion by sampling every 20–30 min, over a 12 h or 24 h period, or by continuous withdrawal (using an indwelling venous catheter and pump) has helped to identify a group of children with low spontaneous GH secretion, despite normal GH responses to pharmacological stimulatory tests (8–10).

The validity and reliability of these tests continue to be debated. Rose *et al.* (11), for example, hold that the measurement of spontaneous GH secretion offers no diagnostic advantage over stimulation tests, while Donaldson *et al.* (12) suggest that pharmacological stimulation underestimates spontaneous GH secretion in approximately 10–20% of children tested, and they further observe that measurement of spontaneous GH secretion for a single night is more reliable for identifying the child with low endogenous GH secretion than is stimulatory testing alone.

The biochemical or pharmacological tests for human GH deficiency can be divided into two groups: drugs that directly stimulate the pituitary and those that stimulate the hypothalamus (Table 1). Direct pituitary stimulation occurs via the use of GH-releasing hormone (GHRH). Pharmacological hypothalamic stimulatory tests can be further subdivided into those substances that increase the release of GHRH and substances that suppress the release of GH-inhibiting hormone (GHIH). stimulatory tests of GHRH are: L-dopa, clonidine, insulin-induced hypoglycemia and glucagon. Substances that

Table 1. Pharmacological stimulatory tests of GH deficiency and mechanisms of action

I. Hypothalamic	Mechanisms of action
A. Stimulatory tests of GHRH	
L-Dopa	Dopaminergic pathways
Clonidine	α_2-Adrenergic receptor
Insulin hypoglycemia	α_2-Adrenergic pathways
Glucagon	α_2-Adrenergic pathways
B. Suppression tests of GHIH	
Arginine	GHIH suppression
Ornithine	GHIH suppression
Galanin	GHIH suppression/cathecholaminergic pathway
Pyridostigmine	GHIH suppression/cholinergic enhancement
II. Pituitary	
GHRH	Direct somatotroph stimulation

suppress the release of GHIH are shown in Table 1; they include arginine, ornithine, galanin and pyridostigmine.

GROWTH HORMONE-RELEASING HORMONE

In 1982, the amino acid sequence of GHRH was characterized (13, 14), and the gene isolated. Recombinant GHRH is now available for human use. The GHRH test allows us to distinguish between pituitary and hypothalamic deficits by assessing pituitary reserves of GH (15–18). Table 2 describes dose and timing of specimens. However, a negative response does not necessarily suggest inability of the pituitary to secrete GH, and some investigators have proposed multiple administrations of GHRH in order to 'prime' the pituitary somatotrophs (19–21).

L-DOPA

L-dopa is an amino acid precursor of dopamine and catecholamines. It causes the release of GH by generating high concentrations of dopamine in the median eminence (22), which then stimulates the secretion of GHRH and increases the serum GH concentration (23–25). The dose and timing of specimens are shown in Table 2 (26).

CLONIDINE

Clonidine is considered to be the most potent stimulator of GH secretion (27). Clonidine acts as a potent α-adrenoceptor-stimulating agent in the brain by binding to both presynaptic and postsynaptic α-adrenoceptors in the hypothalamus or brain (28–32) and stimulates the secretion of GH via increased GHRH (33). The dose and timing of specimens are shown in Table 2. Clonidine is easier, safer and more economical than insulin hypoglycemia (34).

INSULIN HYPOGLYCEMIA

This is one of the most commonly employed tests for evaluating GH deficiency. It has been shown that hypoglycemia promotes the release of various hormones, including GH (35–37). The administration of insulin to the normal child will lower blood sugar levels and generally lead to a compensatory rise in GH secretion. This is an effective test in producing the release of GH into the blood, but it has the serious side effect of hypoglycemia. This may require the hospitalization of the patient for careful monitoring (Table 2).

Table 2. Substances of GH stimulation: dose, timing, side effects

Substance	Dose	Timing (min)	Side effects
L-Dopa	< 15 kg; 125 mg 15–30 kg; 250 mg > 30 kg; 500 mg p.o.	0, 20, 40, 60, 90, 120	Nausea Occasional vomiting
Clonidine	0.15 mg/m^2 p.o.	0, 30, 60, 90, 120	Hypotension drowsiness
Insulin hypoglycemia	0.1 U/kg i.v.	0, 15, 30, 45, 60, 90	Pallor hypoglycemic shock is possible
Glucagon	0.1 mg/kg i.v.	0, 15, 30, 45, 60, 90, 120, 150, 180	Nausea, reactive hypoglycemia is possible
Arginine HCl	0.5 g/kg i.v. over 30 min	0, 30, 60, 90, 120	Nausea, acidosis and hypoglycemia are possible
Ornithine HCl	12 g/m i.v. over 30 min	0, 30, 60	Nausea, acidosis is possible, pallor
Galanin	8 μg/kg per hour i.v.	0, 30, 60	Bitter taste
Pyridostigmine (with GHRH or clonidine)	60 mg p.o. 1 h before GHRH/clonidine	−15, 0, 30, 60, 90, 120, 180	Mild abdominal pain
GHRH	1 μg/kg i.v.	0, 30, 45	Transient facial skin flush

GLUCAGON

Ted AvRuskin was the first to observe that there was a significant response to intramuscular administration of glucagon in a group of children with normal pituitary function (38). As stated previously, glucagon acts directly and indirectly through the α_2-adrenergic pathways, via adrenal gland epinephrine (adrenaline) secretion. Side effects include nausea; reactive hypoglycemia is also possible (Table 2). Of all the stimuli, this is the substance with the most delayed peak time.

ARGININE–ORNITHINE

These are two amino acids which suppress the release of GHIH. Several investigators have demonstrated an adequate peak response to arginine in serum GH concentrations in non-GH-deficient patients, and all have reported a failure of response in GH-deficient children (39–44). This is an effective test, and it has minimal side effects (Table 2).

GALANIN

Recently another neuropeptide—galanin—has been found to be present in significant amounts in the median eminence of the hypothalamus (45, 46). This peptide was found to stimulate GH secretion if administered intravenously in normal subjects (47). Davis *et al.* (48) have reported its mechanism of action to be exerted via catecholaminergic pathways or inhibition of GHIH. Its use has been in conjunction with GHRH or clonidine to potentiate the stimulatory effect of these substances.

PYRIDOSTIGMINE

Pyridostigmine is another stimulatory adjunct used with GHRH to maximize response, especially when the initial GHRH response has been negative (Table 2). It has also been used in conjunction with clonidine (49, 50).

Figure 1 illustrates the effects of the different drugs in the hypothalamus or pituitary. They increase plasma GH (measured by immunoassay).

The question raised is, which of the above stimulatory tests should be used to determine GH deficiency in DS and allow us to identify children that would benefit from hGH treatment? We have presented evidence that children with DS and growth retardation have a biochemical defect in the hypothalamus (2) and a normal pituitary function (64). Tests that directly stimulate the pituitary, like GHRH, will produce a peak serum concentration similar to that found in normal children, making it impossible to diagnose GH deficiency in DS. The clinical value of tests that increase serum GH by suppressing the release of GHIH have not been investigated in growth retardation of neuroendocrine

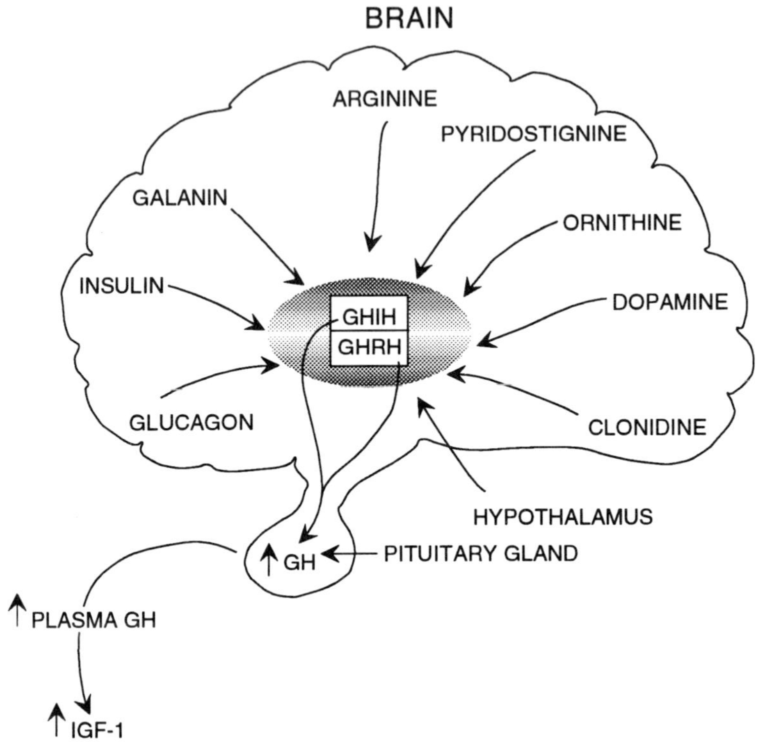

Figure 1. Pharmacological drugs affecting hypothalamic GHRH and GHIH

dysfunction. Thus we can at the present time only use tests that stimulate the hypothalamus to release GHRH in children with DS, and then measure increases in serum GH concentrations. Among the chemicals discussed and shown in Table 2, we have selected L-dopa and clonidine for use in the present study because they have different mechanisms of action and minimal side effects.

METHODS

A total of 26 children with trisomy 21 were evaluated. Their ages ranged from 15 months to 13.9 years (mean age 4.2). Children with CHD were excluded from the study. Twenty-five of the children were cared for by their parents at home and attended infant stimulation programs from the time they were a few weeks of age. The other child is a severely mentally retarded individual (case no. 25) and was institutionalized from early infancy. Another severely mentally retarded individual had panhypopituitarism since birth, and further studies revealed hypoplasia of corpus callosum and septo-optic dysplasia (case no. 5). Thyroid hormonal replacement therapy was instituted in early infancy. The studies were approved by the committee on human experimentation at both the SUNY Health Sciences Center at Brooklyn and the NYS Institute for Basic Research, and informed, written parental consent was obtained before the tests were performed. All tests were performed in duplicate, and the samples from each subject were assayed in the same batch to reduce inter-assay variability.

Height was measured in children over two years of age and length in children below two years of age. Height was measured using a wall-mounted stadiometer. Length was measured with the subject lying on his or her back, with one person holding the child's head. Head circumference (HC) was measured, using a paper tape and selecting the maximal occipitofrontal circumference (51). The average of three consecutive measurements was used as the final value in all growth parameters. The height standard deviation score (SDS) was calculated according to the formula:

$$\text{height SDS} = \frac{\text{height} - \text{mean height of normals at this age}}{\text{SD of height for normals at this age}}$$

Mean height SDS, weight SDS and HC SDS for normal children were obtained from charts developed by Watson andLowery (52). Bone age was determined by obtaining roentgenograms of the left wrist and hand and comparing them with normal roentgenograms for chronological age and sex (53). Serum concentrations of GH, insulin-like growth factor 1 (IGF–1), thyrotropin (TSH), 3,5,3′-triiodothyronine (T_3), thyroxine (T_4), prolactin, luteinizing hormone (LH), follicle-stimulating hormone (FSH), and estradiol or testosterone were determined in all children by standard radioimmunoassay kits. The GH levels from the 12 h IC-GH and 24 h IC-GH studies were performed by Genentech with the Tandem R-hGH immunoradiometric assay kit (Hybritech, San Diego, CA). Intra-assay and inter-assay coefficients of variation for hormonal assays are shown in Table 3.

Table 3. Coefficients of variation for hormonal assays

	Intra-assay (%)	Inter-assay (%)
GH	3.0	9.5
IGF-1	11.8	11.9
TSH	2.7	5.3
T_3	4.9	5.8
T_4	5.0	5.3
Prolactin	3.9	4.1
LH	3.5	6.6
FSH	4.7	9.2
Estradiol	3.9	6.4
Testosterone	4.4	6.4
IC-GH	7.0	6.0

L-Dopa and clonidine stimulation tests for the serum GH concentrations were performed after an overnight fast. L-Dopa was administered orally according to body weight: less than 15 kg body weight received 125 mg, and between 15 and 30 kg body weight received 250 Mg. Blood samples were drawn before and at 20, 40, 60, 90 and 120 min after L-dopa administration. Clonidine was administered orally as a single dose of 0.15 mg/m^2 body surface area. Blood samples for determination of GH levels were drawn immediately prior to the clonidine administrations before and at 30, 60, 90, and 120 min after. A peak GH response to L-dopa or clonidine of > 10 μg/l is considered normal (54–58).

In hospitalized patients, IC-GH was measured during a 24 h or 12 h period by continuous withdrawal of small amounts of blood (2 ml per 30 min periods through an indwelling venous catheter, the withdrawal flow being regulated by a pump). The 12 h IC-GH test was performed in two children younger than 5 years of age and weighing less than 15 kg. Heparin was used in small amounts to flush the catheter at the beginning of the test and at other times, when needed. GH deficiency was defined by a 12 h or 24 h integrated concentration of less than 3.2 μg/1 (59–63).

RESULTS

ANTHROPOMETRIC DATA

Twenty-two out of the 26 (85%) DS children were 2 SD below the mean in height or length for normals. Height SDS of children with DS studied ranged from −1.19 to −5.48 below the mean height for age and sex of control children (Table 4).

Twenty-two of the 26 children were microcephalic, with head circumferences 2 SD below the mean for normals. Head circumference SDS ranged in all DS children studied from −0.40 to −6.600 below the mean for age of control children. The growth rate was below 5 cm per year during the preceding year in the six children older than six years of age and was below the normal corresponding growth rate for children from 1 to 4.5 years (Table 4). Growth velocity was not available for six children. Bone age was retarded in all of the 26 children.

ENDOCRINE DATA

There was no evidence of hypothyroidism in the 25 children with DS studied, as documented by normal serum concentrations of serum T_3, T_4, and TSH. One case had normalized thyroid functions after initial diagnosis of panhypopituitarism and treatment with L-thyroxine. Gonadotropic hormones, estradiol and/or testosterone levels were all within the normal range for prepubertal children.

Table 4. Anthropometric data in children with DS

Patient no.	Age (years, months)	Sex	Height (cm)	Height SDS	Head circumference (cm)	Head circumference SDS	Weight (kg)	Weight SD	Growth velocity (cm/year)	Normal mean height velocity for age[a]
1	1,3	M	74.0	−2.09	46.5	−0.40	9.8	−0.90	12	25
2	1,4	M	72.5	−2.68	43.5	−3.29	9.8	−0.21	8	25
3	1,4	F	71.0	−2.20	42.0	−4.54	7.6	−2.64	8	25
4	1,4	M	73.0	−2.49	44.0	−2.90	8.4	−3.16	–	25
5	1,5	M	65.5	−5.43	42.0	−4.54	6.7	−4.58	7	25
6	1,5	M	76.0	−1.31	44.5	−2.45	11.0	−1.00	9	25
7	1,6	F	75.0	−1.85	43.5	−4.00	9.2	−1.50	7	25
8	1,7	M	79.5	−1.19	46.5	−1.58	11.0	−1.00	10	25
9	1,10	M	81.5	−2.03	46.5	−2.00	11.0	−1.57	7	10
10	1,10	F	73.0	−3.28	43.0	−4.75	8.3	−2.57	–	10
11	2,11	M	86.0	−3.06	47.0	−2.50	11.6	−2.25	5	10
12	3,2	M	88.0	−2.47	45.5	−3.75	12.7	−1.56	6	7–8
13	3,3	F	84.0	−3.16	42.0	−6.60	11.5	−1.77	5	7–8
14	3,7	F	89.2	−2.78	46.7	−2.90	14.7	−0.63	4	7–8
15	3,9	M	87.6	−2.56	47.0	−2.85	14.5	−0.56	–	6–7
16	4,4	M	99.0	−1.31	47.0	−2.90	16.4	−0.45	4	6–7
17	4,5	F	92.0	−3.62	47.0	−2.76	13.0	−2.12	5	6–7
18	4,11	M	90.5	−3.68	46.0	−3.43	16.6	−0.78	–	5–6
19	5,4	M	101.1	−2.77	46.8	−3.00	15.0	−1.42	–	5–6
20	5,10	F	100.0	−2.85	47.0	−2.85	19.0	−0.49	–	5
21	6,0	M	101.0	−3.79	48.5	−1.92	18.0	−1.50	4	5
22	6,6	M	110.0	−2.37	50.5	−0.60	22.0	−0.51	4	5
23	6,6	F	106.0	−4.78	47.5	−2.70	20.0	−0.89	2	5
24	9,0	M	111.0	−4.78	46.7	−3.60	19.0	−2.45	4	5
25	11,0	M	117.5	−4.65	49.0	−2.77	24.5	−1.96	4	5
26	13,9	F	131.1	−5.48	49.5	−2.77	26.0	−3.39	4	5

[a]Obtained from Lifshitz F, Zandsberg S, Disorders of growth. In *Pediatric Endocrinology*, edited by Lifshitz F, Marcel Dekker, New York, 1985; pp. 11–12.

Table 5. Responses to L-dopa and clonidine testing of children with and without DS

Groups		L-dopa (basal)	L-dopa (peak)	L-dopa (peak – basal)	Clonidine (basal)	Clonidine (peak)	Clonidine (peak – basal)
Non-DS controls		0.400	11.800	11.400	3.100	20.600	17.500
		1.000	12.900	11.900	0.800	19.000	18.200
		0.800	30.200	29.400	0.900	26.800	25.900
		1.800	14.700	12.900	1.900	10.400	8.500
		21.800	18.700	−3.100	8.300	28.200	19.900
		3.900	25.700	21.800	5.600	24.700	19.100
		12.900	14.400	1.500	1.700	14.000	12.300
	Mean	6.08	18.34	12.25	3.18	20.52	17.34
	SD	(8.19)	(7.02)	(11.08)	(2.79)	(6.62)	(5.58)
DS low-low		5.400	4.400	−1.000	0.800	2.500	1.700
		9.300	9.900	−0.600	3.900	9.500	5.600
		1.300	1.500	−0.200	1.000	1.600	0.600
		2.100	2.000	−0.100	1.400	6.000	4.600
		1.600	7.300	5.700	1.800	3.600	1.800
		4.200	7.300	3.100	3.800	9.700	5.900
		5.600	7.600	2.000	23.000	7.000	−16.000
		1.500	1.700	0.200	0.900	4.500	3.600
		5.600	3.800	−1.800	1.500	7.900	6.400
	Mean	4.06	5.05	0.98	4.23	5.81	1.57
	SD	(2.69)	(3.06)	(2.29)	(7.13)	(2.95)	(6.90)
DS low-normal		1.200	14.100	12.900	2.700	6.100	3.400
		6.500	12.800	6.300	2.400	6.500	4.100
		5.300	29.300	24.000	26.100	10.000	−16.100
		3.400	5.900	2.500	3.000	14.800	11.800
		2.200	19.800	17.600	2.700	8.900	6.200
		8.500	13.900	5.400	1.800	4.000	2.200
		14.900	4.900	−10.000	2.400	10.200	7.800
		8.600	8.300	−0.300	0.800	15.300	14.500
		1.500	10.400	8.900	1.300	3.000	1.700
		1.500	23.500	22.000	18.500	5.500	−13.000
		2.900	2.800	−0.100	1.100	19.000	17.900
		0.800	6.700	5.900	0.800	11.800	11.000
		1.500	28.100	26.600	4.300	1.500	−2.800
		11.400	4.600	6.800	0.800	35.500	34.700
		1.100	64.800	63.700	4.400	9.800	5.400
	Mean	5.38	15.92	11.90	9.83	10.68	5.92
	SD	(4.90)	(15.54)	(17.94)	(16.66)	(8.30)	(12.15)
DS normal-normal		14.900	4.900	−10.000	25.700	14.400	−11.300

Nine of the 25 children studied (36%) had abnormally low peak increase in serum GH response to both L-dopa and clonidine testing (peak GH serum concentration < 10 ng/ml). (This is the 'low-low' in Table 5). Sixteen children (64%) showed a disparity between L-dopa and clonidine testing, i.e. they were low on either L-dopa or clonidine in Table 5. This is the 'low-normal' group in Table 5. Of these 16 children, nine (56%) had L-dopa peak increase in serum GH concentration above 10 ng/ml and eight (50%) patients had clonidine peak increase in serum GH concentration above 10 ng/ml.

One out of the 25 children with DS studied responded to both stimulatory tests with a peak GH serum concentration > 10 ng/ml. If there was more than one case, we would call this the normal response group. Of the 16 patients that showed a disparity between stimulatory tests, two older than five years of age had a 24 h IC-GH in the range of those patients with GH deficiency (Table 5). Figure 2 compares the 24 h IC-GH of two children with DS to one non-DS control patient. The two patients with DS had a 24 h integrated GH deficiency.

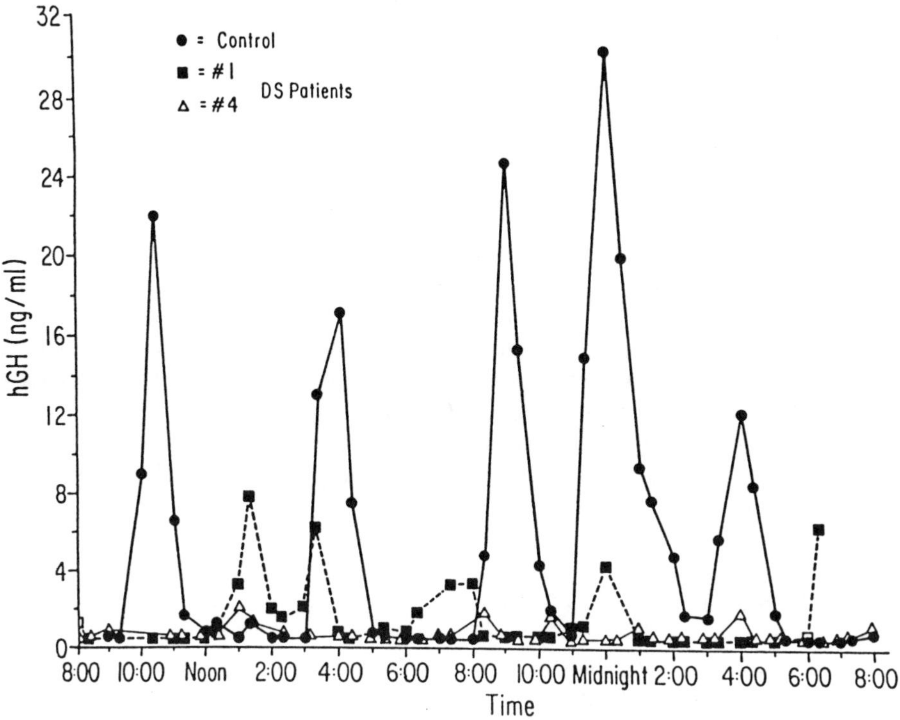

Figure 2. The 24 h IC-GH in two children with DS and one normal control child

DISCUSSION

By stimulating the hypothalamus to produce GHRH, L-dopa and clonidine induce the pituitary to release measurable amounts of GH into the blood. This makes both L-dopa and clonidine valuable in the determination of GH deficiency in children with DS. They have different mechanisms of action, and in this study they were found to be uncorrelated statistically ($\rho = -0.12$). In the study, three different types of responses to L-dopa and clonidine were found. Children with DS and short stature in the 'low-low' group showed low responses of serum GH concentrations to both L-dopa and clonidine stimulation; these children have GH deficiency, and we believe recombinant human growth hormone (rhGH) treatment is indicated. Using the Kruskal–Wallis rank order analysis of variance, the 'low-low' group was significantly different statistically from the non-DS control group ($p = 0.035$ and 0.001 for L-dopa and clonidine, respectively). The second or 'low-normal' group is characterized by a low measurement on one test and a normal one on the other. The difference between the 'low-low' group and 'low-normal' group is significant for L-dopa ($p = 0.049$) but not for clonidine ($p = 0.283$). It is difficult to reach a conclusion on the use of rhGH treatment with children in this group. The authors had previously reported that the limited or disparate responses to stimuli in children with DS may reflect a state of diminished GH secretion This disparity of responses to L-dopa and clonidine suggests to us that children with DS have a GH 'inadequacy' secondary to the different abnormalities in the stimulatory effect of hypothalamic centers in the release of pituitary GH (2). The authors feel that treatment with hGH is also indicated in this group. It also should be noted that one child with DS and short stature in this study had a normal response to each test (this case would be part of a 'normal-normal' group); hGH treatment is not indicated for children with normal responses to both tests. Twelve-hour or 24 h IC-GH concentrations in the patients that had the test done also supported GH deficiency. A normal GH response to GHRH in DS indicates normal pituitary (normal somatotrope response to GHRH) and supports hypothalamic dysfunction in DS as a cause of GH inadequacy. The authors acknowledge that not all DS children have GH deficiency, but the blunted or diminished responses may be a contributory factor in the cause of the associated short stature that is well documented in these children.

REFERENCES

1. Cronk C, Crocker AC, Pueschel SM *et al.* Growth charts for children with Down's Syndrome: 1 month to 18 years of age. *Pediatrics* 1988; **81:** 102–110.
2. Castells C, Torrado C, Bastian W *et al.* Growth hormone deficiency in Down's syndrome children. *J Intell Disability Res* 1992; **36:** 29–43.

3. Buckler JMH. Exercise as a screening test for growth hormone release. *Acta Endocrinol.* 1972; **69:** 219–229.
4. Keenan BS, Killmer LB, Sode J. Growth hormone response to exercise: a test of pituitary function in children. *Pediatrics* 1972; **50:** 760–764.
5. Okada Y, Hikita T, Ishitobi K *et al.* Human growth hormone secretion after exercise and oral glucose administration in patients with short stature. *J Clin Endocrinol Metab* 1972; **34:** 1055–1058.
6. Richards GE, Cavallo A, Meyer WJ. Diagnostic validity of 12 hour integrated concentration of growth hormone. *Am J Dis Child* 1987; **141:** 553–555.
7. Tassoni P, Cacciari E, Cau M *et al.* Variability of growth hormone response to pharmacological and sleep tests performed twice in short children. *J Clin Endocrinol Metab* 1990; **71:** 230–234.
8. Spiliotis BE, August GP, Hung W *et al.* Growth hormone neurosecretory dysfunction: a treatable cause of short stature. *JAMA* 1984; **251:** 2223–2230.
9. Zadik Z, Chalew SA, Raiti S *et al.* Do short children secrete insufficient growth hormone? *Pediatrics* 1985; **76:** 355–360.
10. Bercu BB, Shulman D, Root AW *et al.* Growth hormone (GH) provocative testing frequently does not reflect endogenous GH secretion. *J Clin Endocrinol Metab* 1986; **63:** 709–716.
11. Rose SR, Ross JL, Uriarte M *et al.* The advantage of measuring stimulated as compared with spontaneous growth hormone levels in the diagnosis of growth hormone deficiency. *N Engl J Med* 1988; **319:** 201–207.
12. Donaldson DL, Pan F, Hollowell JG *et al.* Reliability of stimulated and spontaneous growth hormone (GH) levels for identifying the child with low GH secretion. *J Clin Endocrinol Metab* 1991; **72:** 647–652.
13. Guillemin R, Brazeau P, Bohlen P *et al.* Growth hormone-releasing factor from a human pancreatic tumor that caused acromegaly. *Science* 1982; **218:** 585–587.
14. Rivier J, Spiess J, Thorner M *et al.* Characterization of a growth hormone-releasing factor from a human pancreatic islet tumor. *Nature* 1982; **300:** 276–278.
15. Rosenthal SM, Schriock EA, Kaplan SL *et al.* Synthetic human growth hormone-releasing factor (hp GRF-1-44-NH2) stimulates growth hormone secretion in normal men. *J Clin Endocrinol Metab* 1983; **57:** 677–679.
16. Thorner MO, Spiess J, Vance ML *et al.* Human pancreatic growth hormone-releasing factor selectively stimulates grow hormone secretion in man. *Lancet* 1983; **ii:** 24–28.
17. Gelato MC, Pescovitz O, Cassorla F *et al.* Effects of growth hormone-releasing factor in man. *J Clin Endocrinol Metab* 1983; **57:** 674–676.
18. Albini CH, Reiter EO, Mills BJ *et al.* Diagnostic value of the growth hormone-releasing factor stimulation test. *Clin Pharmacol Ther* 1988; **43:** 696–700.
19. Schriock EA, Lustiga RH, Rosenthal SM *et al.* Effect of growth hormone (GH)-releasing hormone (GRH) on plasma GH in relation to magnitude and duration of GH deficiency in 26 children and adults with isolated GH deficiencies. *J Clin Endocrinol Metab* 1984; **58:** 1043–1049.
20. Rogol AD, Blizzard RM, Johanson AJ *et al.* Growth hormone release in response to human pancreatic tumor growth hormone-releasing hormone in 40 children with short stature. *J Clin Endocrinol Metab* 1984; **59:** 580–585.
21. Grossman A, Savage MO, Blacklay A *et al.* The use of growth hormone-releasing hormone in the diagnosis and treatment of short stature. *Hormone Research* 1985; **22:** 52–57.
22. Wurtman R. Brain catecholamines and the control of secretion from the anterior pituitary gland. In *Hypophysiotropic hormones of the Hypothalamus: Assay and Chemistry*, edited by Meites J, Williams & Wilkins, Baltimore, 1970; pp. 184–194.

23. Donnadiew M, Evam-Brion D, Tonan MC *et al.* Variations of plasma growth hormone (GH) releasing factor levels during GH stimulation tests in children. *J Clin Endocrinol Metab* 1985; **60:** 1132.
24. Chihara K, Kashio Y, Kita T *et al.* L-Dopa stimulates release of hypothalamic growth hormone releasing hormone in humans. *J Clin Endocrinol Metab* 1986; **62:** 466–473.
25. Tapanninen P, Knip M, Lautala P *et al.* Variable plasma growth hormone (GH)-releasing hormone and GH responses to clonidine, L-dopa, and insulin in normal men. *J Clin Endocrinol Metab* 1988; **67:** 845–849.
26. Weldon VV, Gupta SK, Haymond MW. The use of L-dopa in the diagnosis of hyposomatotropism in children. *J Clin Endocrinol Metab* 1973; **36:** 42.
27. Shulman D, Zamanillo J, Lowitt S. Effect of clonidine upon anterior pituitary function and plasma catecholamine concentrations in short children and adolescents. *J Pediatr Endocrinol* 1985; **4:** 211–216.
28. Schmitt H, Schmitt H, Fernard S. Evidence for an alpha-sympathomimetic component in the effects of catapresan on vasomotor centers: antagonism of piperoxan. *Eur J Pharmacol* 1971; **14:** 98–100.
29. Reid JL, Tangri KK, Wing LMH. The central hypotensive action of clonidine and propranolol in animals and man. *Prog Brain Res* 1977; **47:** 369–384.
30. Cavero I, Roach AG. The effects of prazosin on the clonidine induced hypotension and bradycardia in rats and sedation in chicks. *Br J Pharmacol* 1978; **62:** 468–469.
31. Drew GM, Gower AJ, Marriott AS. Alpha 2-adrenoceptors mediate clonidine-induced sedation in the rat. *Br J Pharmacol* 1979; **67:** 133–141.
32. Timmermans PB, Lam E, Van Zwieten PA. The interaction between prazosin and clonidine at alpha-adrenoceptors in rats and cats. *Eur J Pharmacol* 1979; **55:** 57–66.
33. Miki N, Ono M, Shizume K. Evidence that opiatergic and adrenergic mechanisms stimulate rat growth hormone release via growth hormone releasing factor (GRF). *Endocrinology* 1984; **114:** 1950–1952.
34. Singh SK, Hatwal A, Agrawal JK *et al.* Oral clonidine: an effective growth hormone provocative test. *Indian Pediatr* 1989; **26:** 1007–1009.
35. Roth J, Glick SM, Yalow RS *et al.* Hypoglycemia: a potent stimulus to secretion of growth hormone. *Science* 1963; **140:** 987.
36. Kaplan SL, Abrams CAL, Bell JJ *et al.* Serum growth hormone response to insulin-induced hypoglycemia in disorders of growth. *J Pediatr* 1965; **67:** 956.
37. Kaplan SL, Abrams CAL, Bell JJ *et al.* Growth and growth hormone: I. Changes in serum level of growth hormone following hypoglycemia in 134 children with growth retardation. *Pediatr Res* 1968; **2:** 43.
38. AvRuskin TW, Crigler JF Jr, Sonksen PH, Soeldner JS. Stimulation tests of growth hormone secretion. In *Recent Advances in Endocrinology Proceedings*, edited by Matter E, Matter G, James VHT. Pan American Congress of Endocrinology (7th, 1970, Sao Paulo). Amsterdam Excerpta Medica 1971 International Congress Series No. 238 pp. 395–402.
39. Frohman LA, Aceto T Jr, MacGillivray MG. Studies of growth hormone secretion in children: normal, hypopituitary and constitutionally delayed. *J Clin Endocrinol* 1967; **27:** 1409.
40. Haiti S, Davis WT, Blizzard RM. A comparison of the effects of insulin hypoglycemia and arginine infusion on release of human growth hormone. *Lancet* 1967; **ii:** 1182.
41. Parker ML, Hammond JM, Daughaday WH. The arginine provocative test: an aid in the diagnosis of hyposomatotropism. *J Clin Endocrinol* 1967; **37:** 1129.

42. Knopf RF, Conn JW, Fajans SS *et al.* Plasma growth hormone response to intravenous administration of amino acids. *J Clin Endocrinol* 1965; **25:** 1140.
43. Merimee TJ, Lillicrap DA, Rabinowitz D. Effect of arginine on serum levels of human growth hormone. *Lancet* 1965; **ii:** 668.
44. Root AW, Saenz-Rodriguez C, Bongiovanni AM *et al.* The effect of arginine infusion on plasma growth hormone and insulin in children. *J Pediar* 1969; **74:** 187.
45. Tatemoto K, Rokaeus A, Jornvall H *et al.* Galanin: a novel biological active peptide from porcine intestine. *FEBS Lett* 1983; **164:** 124.
46. Rokaeus A, Melander T, Hokfeit T *et al.* A galanin-like peptide in the central nervous system and intestine of the rat. *Neurosci Lett* 1984; **47:** 161.
47. Bauer FE, Ginsberg L, Venetikou M *et al.* Distribution of galanin immunoreactivity in the genitourinary tract of man and rat. *Peptides* 1986; **7:** 5.
48. Davis TME, Burrin JM, Bloom SR. Growth hormone (GH) release in response to GH-releasing hormone in man is 3-fold enhanced by galanin. *J Clin Endocrinol Metab* 1987; **65:** 1248–1252.
49. Ghigo E, Mazza E, Imperiale E *et al.* The enhancement of cholinergic tone by pyridostigmine promotes both basal and growth hormone releasing hormone induced GH secretion in children of short stature. *J Clin Endocrinol Metab* 1987; **65:** 452.
50. Ghigo E, Goffi S, Mazza E *et al.* Acute administration of pyridostigmine and clonidine has an additive stimulatory effect on GH release in normal children. *J Endocrinol Invest* 1989; **12:** 99.
51. Hoekelman RA. The pediatric physical examination. In *A Guide to Physical Examination*, edited by Bates B, Lippincott, New York, 1974; pp. 320–362.
52. Watson EH, Lowery GH. *Growth and Development of Children*, 4th ed, Year Book Medical Publishers, Chicago, 1962.
53. Greulich WW, Pyle SI. *Radiographic Atlas of Skeletal Development of the Hand and Wrist*, 2nd edn, Stanford University Press, Stanford, 1959.
54. Gil-Ad I, Topper E, Laron Z. Oral clonidine as a growth hormone stimulation test. *Lancet* 1979; **ii:** 278–279.
55. Lanes R, Hurtado E. Oral clonidine: an effective growth hormone-releasing agent in prepubertal subjects. *J Pediatr* 1982; **100:** 710–714.
56. Delitala G, Maioli M, Pacifico A *et al.* Cholinergic receptor control mechanisms for L-dopa, apomorphine, and clonidine-induced growth hormone secretion in man. *J Clin Endocrinol Metab* 1983; **57:** 1145–1149.
57. Checkley SH, Slade AP, Shur E. Growth hormone and other responses to clonidine in patients with endogenous depression. *Br J Psychiatry* 1981; **138:** 51–55.
58. Spilliotis BE, August GP, Hung W *et al.* Growth hormone neurosecretory dysfunction, a treatable cause of short stature. *JAMA* 1984; **251:** 2223–2230.
59. Zadik Z, Chalew SA, Raiti S *et al.* Do short children secrete insufficient growth hormone? *Pediatrics* 1985; **76:** 355–360.
60. Frasier SD, Costin G, Lippe BM *et al.* A dose–response curve for human growth hormone. *J Clin Endocrinol Metab* 1981; **53:** 1213–1217.
61. Zadik Z, Chalew SA, McCarter RJ *et al.* The influence of age on the 24-hour integrated concentration of growth hormone in normal individuals. *J Clin Endocrinol Metab* 1985; **60:** 513–516.
62. Kowarski AA. Bioinactive growth hormone and secondary growth hormone deficiency. In *Human Growth Hormone*, edited by Raiti S, Plenum Press, Baltimore, 1986; pp. 127–134.
63. Plotnick LP, Thompson RG, Kowarski AA *et al.* Circadian variation of integrated concentration of growth hormone in children and adults. *J Clin Endocrinol Metab* 1975; **40:** 240–247.

Discussion of Presentation

CARMEN TORRADO
(Edited by D.A. Snider and Salvador Castells)

P. Wilton: You mentioned that there is always a problem evaluating various kinds of stimulation tests in overweight children. As many of them are did you try to make any comparisons to see if the non-overweight children had a different response on the different tests from the non-overweight children with Down's syndrome?

C. Torrado: No, but what we did find was that their weights were closer to normal than their height, but none of these children were obese.

V. Sara: I have two questions concerning the IGF-1 levels you presented. Did you see, or fail to see, a rise in IGF-1 levels during puberty in these children?

C. Torrado: The only girl studied of pubertal age was 13 years 9 months, and had a Tanner 2 (was at the onset of puberty). All of the other children were prepubertal.

V. Sara: It would be interesting to see because the IGF-1 levels increase during this time and that was the really striking feature of the study we did—that they didn't increase during puberty. How did you determine the IGF-1 levels? Did you separate binding proteins first in these assays?

C. Torrado: Somatomedin-C was determined using a standard radioimmunoassay kit (from Nichols Institute Diagnostics, San Juan Capistrano, California). This method includes the binding protein in the total measurement of somatomedin C. At present we are using an immunoassay that measures IGF-l after acid extraction and purification on octadecasilyl silica cartridges and does not measure binding proteins.

V. Sara: I think, then, caution has to be taken, because as I am sure you are aware, there are difficulties in the interpretation of these results.

G. Merriam: I want to follow up on the question, that overweight may significantly modify the results. Earlier there were some studies showing major reductions in spontaneous growth hormone secretion and response to provocative testing in patients with morbid obesity. In a collaborative study of a group of aging individuals we did with the National Institute of Aging a negative correlation was found between body mass index (BMI) and growth hormone responses to provocative testing, even within the normal range, and I think it potentially very useful for you to try to plot your growth hormone results against the BMI of these children, because if there is some heterogeneity you may in fact find that much of the variance in your growth hormone data could be accounted for by high BMI, rather than a specific defect in growth hormone secretion. As these children move from one type of treatment setting to another and experience changes in weight, retesting their growth hormone status, if they are able to obtain a more normal body weight for their size we might find an increment in spontaneous as well as stimulated growth hormone secretion.

C. Torrado: We certainly should look into that.

N. Stahnke: If patients with Down's syndrome really suffer from hGH deficiency, in general we would expect much more normal auxological data, for instance severely stunted growth, severely retarded bone age—and what is most puzzling, you found hGH deficiency in your assays, but IGF-1 was normal; this is really not hGH deficiency. I am not convinced that these patients have this endocrine dysfunction in general, for we know that so many patients with a GH deficiency depart clearly from the picture you showed us.

C. Torrado: Classical growth hormone deficiency in this group is relatively rare. What we are seeing is growth hormone inadequacy or partial growth hormone deficiency. Children are able to secrete some growth hormone, but not enough for normal growth. Indeed, it is quite debatable today in the growth hormone literature whether these children with growth hormone inadequacy should be treated. The problem is, how do we go about making the diagnosis of inadequate growth hormone in short children, whether or not they have Down's syndrome? This is a challenge, because we do not have one significant test for growth hormone deficiency and we have to employ all of these hypothalamic stimulation tests. Maybe we need to study the hypothalamic axis, using all of the different stimulators—the clonidine, the L-dopa, as well as other tests—to see what the different responses are in these patients. Inadequate or partial growth hormone deficiency in these patients is not so easily diagnosed.

6 Thyroid Dysfunction and Growth Retardation in Down's Syndrome

TERESA SHARAV
Jerusalem Child Development Center, Jerusalem, Israel

INTRODUCTION

The history of the connection between hypothyroidism and Down's syndrome (DS) initially arose from the observation of strikingly similar clinical features between hypothyroid infants and DS. Benda's finding (1) in the 1960s of a small-sized thyroid gland with underdeveloped histology seemed to provide pathological evidence for the clinical observation. Hence the failure of routine thyroid treatment to improve either intellectual functioning or growth of the DS as demonstrated by Koch (2) was certainly disappointing. Koch's study population was that of DS infants and children from the age of one year to three years. Following Koch's study of 1965 there were a group of studies in the 1970s (3–5) that described thyroid function in the adolescent and adult DS subject. These studies showed that there was an increased prevalence of abnormal thyroid tests and hypothyroidism in this population. The type of hypothyroidism mainly described was that of the autoimmune type with increased autoantibodies. Further epidemiological studies showed that there was an increased prevalence of autoantibodies in mothers and other female relatives of DS individuals (6). With the advent of neonatal population screening in the 1980s hypothyroidism was described as occuring at the rate of 1 : 141 in DS (7), with a rate of 1 : 377 of what was defined as transient hypothyroidism. The rate of hypothyroidism in the general population is approximately 1 : 4000. Also about this same period of time studies of thyroid function in the DS (8, 9) described an increased incidence of higher than expected basal thyrotropin (TSH) levels in the presence of normal thyroxine (T_4) and triiodothyronine (T_3) levels. This is now termed thyroid dysfunction and has been found to occur throughout the age spectrum in the DS population. In infancy, unlike the hypothyroidism of adolescence or adulthood, there is not an association with thyroid autoantibodies (9, 10). Thyroid dysfunction has been variously defined as subclinical hypothyroidism, biochemical hypothyroidism, preclinical hypothyroidism and decreased thyroid reserve. The consequences of the finding of thyroid dysfunction as expressed by a high TSH level on physical growth and intellectual development are not yet clear.

Growth Hormone Treatment in Down's Syndrome. Edited by S. Castells and K.E. Wisniewski

The IQ levels of DS subjects between the ages of 3 and 21 years with high basal TSH but normal T_4 values did not differ from IQ levels of DS subjects with normal basal TSH levels (8). Treating DS individuals with thyroid supplement between the ages of 2 and 52 years with high TSH levels for a period of two to three months showed no change in intellectual function (11). However, these are not unexpected findings since it is well described that the consequences of thyroid deficiency are most marked in early infancy, and for thyroid supplement to be effective it has to be given within at least the first three months of life (12).

One of the key problems of investigating thyroid function in infancy is having clear criteria of what is considered normal thyroid function. At birth and during the immediate neonatal period there is a TSH surge of up to 70 mU/l, which declines to 10 mU/l in the next two days. T_4 and T_3 concentrations increase over the next two days, peaking at two to three days, and decrease slowly over the next weeks of life. By the age of one to two months, the values should approximate those of adults. Taking into account the variations of thyroid hormone levels with age, certain criteria were established by Fisher (13). However,during the early period of transition from the fetal to adult pattern of secretion interpretation of results still continues to provide difficulties. Since increased levels of TSH are accepted as being the most sensitive indicator of possible thyroid dysfunction, in instances when the picture of thyroid function is not clear it is considered helpful to do the thyrotropin-releasing hormone (TRH) stimulation test which detects even mild degrees of thyroid failure. Hence we used the TRH test to help elucidate thyroid dysfunction in DS.

MATERIALS AND METHODS

To attempt to clarify the picture of thyroid function and dysfunction in the DS population a longitudinal study of the biochemistry of thyroid function from the age of one month to early childhood was performed. Two further studies on the interrelationship between thyroid dysfunction and growth are described: the first is a retrospective study on growth in infants from four months to four years with elevated levels of TSH and the second investigates the effect of thyroid supplementation in infants with elevated TSH values.

BIOCHEMISTRY OF THYROID FUNCTION IN DOWN'S SYNDROME

Forty-seven infants and children with DS were included in this longitudinal study. Chromosomal analysis showed a trisomy 21 was present in 45 children, one was a translocation and one was a mosaic. Initially the patients were included as they came for routine follow-up at the Jerusalem Child Development Center, and ranged in age from one month to seven years. However, as new infants were born they were studied within two months of birth, so consequently there were more infants in the youngest age category. All the infants included in the study were born within three weeks of term, were in good health, and did not have clinical goiter. Subjects were examined after an overnight fast or, in infants, 4 h after their early morning feed. The TRH stimulation test was performed by injecting an intravenous bolus of 4 mg/kg synthetic TRH (Roche); serum TSH was determined before and at 15, 30, and 60 min post-injection. The higher of the 15 or 30 min value was designated as the peak value. Total T_4 and T_3 were measured in unextracted serum by radioimmunoassay (RIA) (Seralute Ames Co., Elkhart, IN; T_3 RIA Kit, Amersham, UK). Serum TSH was measured by RIA (14) with a sensitivity of 0.5–1.0 mU/ml. Antibody titers to thyroglobulin and thyroidal microsomal antibody were measured by immunofluorescence. The normal adult ranges for the laboratory are: T_4, 59–154 nmol/l, T_3, 80–200 ng/dl; TSH, 0.5–6 mU/ml. A peak TSH response of above 20 mU/ml is classified as exaggerated. The criteria of Fisher were used to diagnose hypothyroidism as a function of age. The values are as follows: from 1 to 12 months, T_4, 80–198 nmol/l, T_3, 103–219 ng/dl, TSH 0.5–6.5 mU/ml; from one to six years, T_4, 68–198 nmol/l, T_3, 100–224 ng/dl, TSH 0.6–6.3 mU/ml; and from 7 to 12 years, T_4, 62–160 nmol/l, T_3, 83–211 ng/dl, TSH, 0.6–6.3 mU/ml. The TSH response to TRH stimulation in DS was compared to controls. An age-matched control group of 22 infants and children—11 males and 11 females—was obtained from the Pediatric Endocrinology Department. These were infants and children who during the period of the study were investigated for various clinical reasons by the TRH test and in whom no subsequent biological or endocrine abnormality was found. Thirty-three of the DS subjects who had only one TRH test performed were compared to the controls. The DS children and controls were grouped according to age: under one year, one to three years and above three years. A subgroup of 14 DS subjects who had an initial exaggerated response to TRH stimulation was analyzed separately. All 14 patients had a repeat TRH test after approximately one year and six patients had three annual repeat TRH tests performed. Eighteen of the tests were performed under the age of one year, 11 between the age of one and three years and five over the age of three years. For statistical analysis, a repeated measures analysis of variance was done to examine differences between DS and controls at varying age groups (15, 16). For the group of 14 DS subjects who had follow-up repeat TRH tests, Duncan's multiple range test was used. This was used to partition the effects of age, within-subject variation and across-subject variation in TSH.

GROWTH IN DOWN'S SYNDROME INFANTS AND CHILDREN WITH ELEVATED THYROTROPIN VALUES

Ninety-four home-reared DS subjects under the age of four years—patients of the DS clinic at the University of Baltimore—were included in a retrospective study of thyroid function (17). All three types of DS—trisomy, translocation and mosaic—were included. The subjects had routine T_4 and TSH analysis and were screened for the presence of thyroid antibodies. The subjects were subgrouped according to a TSH value of either a TSH higher or lower than 5.7 mU/l, i.e. a high TSH (HTSH) and normal TSH (NTSH) group. Two age groups were defined between four months and two years, and between two and four years. The mean age in the first group was 11.1 months for the HTSH group and 13.3 months for the NTSH group. The mean age in the two- to four-year group was 37.2 months for the HTSH group and 34.4 months for the NTSH group. The T_4 values for the infants under four years were significantly higher in the patients with HTSH than NTSH (143 versus 121 nmol/l respectively, $p < 0.001$). Growth parameters were available in 82 infants. The correlations between blood TSH and growth were studied using polynomal regression analysis. The best-fitting curve was selected according to the significance of correlation expressed as '*t*' value.

EFFECT ON GROWTH OF THYROID SUPPLEMENT IN DOWN'S SYNDROME INFANTS WITH HIGH THYROTROPIN VALUES

To date we have the longitudinal growth parameters for the first three years of life of 36 DS infants—17 females and 19 males. None of these infants suffer from a heart defect or other congenital abnormality. Infants were examined within one month of birth and administered the TRH test by two and a half months at the latest. Growth of all infants was measured at each examination: head circumference using a tape, height or length on a measuring board and weight on an electronic scale. Following the TRH stimulation test the subjects were divided into an HTSH and NTSH group using the basal TSH value appropriate for age according to the criteria of Fisher. The group of HTSH DS infants with a high basal TSH result were further divided into a treatment and non-treatment group. Subjects were alternately assigned to each group. Treatment was started by the age of three months at the latest and was with L-thyroxine 8 μg/kg. Follow-up consisted of a repeat TRH test as described in the study of biochemical function, i.e. at about one, two and three years of age. All growth parameters were again measured at this time as in the routine examinations at one and six months. Those DS patients receiving thyroid supplement continued to receive it until they reached a TRH test result with both basal and peak TSH values within normal limits. The DS patients with an NTSH had no further TRH tests, just annual basal thyroid tests. No tests showed conversion from normal TSH to high TSH and all T_4 and T_3 values were within the normal range. Statistical analysis was performed using Student's *t*-test.

RESULTS

BIOCHEMISTRY OF THYROID FUNCTION IN DOWN'S SYNDROME INFANTS AND CHILDREN

Mean T_4 and T_3 values were within the normal range in both DS and control groups (Table 1). All thyroid antibody studies were negative. Using the criteria of Fisher (13), none of the DS infants had T_4 levels sufficiently low to be designated clearly as hypothyroid. Table 1 shows T_4 values in relation to basal, peak and 60 min TSH levels in subjects with DS and controls. Repeated measures ANOVA showed that the mean basal TSH levels of subjects with DS were significantly higher than those of age-matched controls ($f = 4.1$, $p < 0.05$). The peak TSH response was significantly higher in the DS population than in the control group ($f = 11.9$, $p < 0.001$). The mean 60 min TSH level was also significantly higher in the subjects with DS than in the controls ($f = 4.0$, $p < 0.05$). This difference between DS subjects and controls held true for all age categories, even though the basal and peak TSH levels in DS declined to what is considered normal levels for age. The subgroup of 14 subjects with DS who initially had an exaggerated TSH response to TRH was studied longitudinally.

Table 1. Means and (±) standard deviations of T_4 and levels of TSH response to TRH injection in DS and controls by age groups

	n	T_4 (nmol/l)	TSH (mU/l) 0 min	TSH (mU/l) Peak (15–30 min)	TSH (mU/l) 60 min
Overall					
Down's	33	126 ± 35	7.3 ± 6.9	25.8 ± 11.4	18.2 ± 11.4
Control	22	142 ± 19	2.5 ± 1.8	14.7 ± 5.7	10.8 ± 5.4
Under 1 year					
Down's	19	126 ± 42	10.2 ± 7.7	30.5 ± 11.3	23.3 ± 13.5
Control	7	138 ± 19	4.0 ± 2.3	18.0 ± 4.2	15.5 ± 5.3
1–3 years					
Down's	7	127 ± 23	4.2 ± 3.9	23.9 ± 8.1	15.3 ± 4.4
Control	6	133 ± 19	2.3 ± 1.1	17.5 ± 5.4	10.9 ± 3.9
3–7 years					
Down's	7	116 ± 22	2.7 ± 0.6	14.9 ± 4.8	9.6 ± 3.6
Control	9	153 ± 21	1.3 ± 0.8	9.0 ± 2.4	5.9 ± 1.7

ANOVA:
T_0 $f = 4.1$ $p < 0.05$
T_{peak} $f = 11.9$ $p < 0.001$
T_{60} $f = 4.0$ $p < 0.05$.

Table 2. Means and (±) standard deviations of basal and peak TSH levels in a group of 14 DS subjects studied longitudinally

	<1 year	1–3 years	3–7 years
No. of test	11	18	5
Basal TSH (mU/ml)	8.1 ± 4.9	6.8 ± 3.9	2.1 ± 1.0
Peak TSH (mU/ml)	35.9 ± 7.1	36.4 ± 10.1	24.3 ± 13.5
	A	A	B

Means with the same letter are not significantly different.
Duncan's multiple range test:
Basal TSH $f = 3.45$ $p < 0.05$
Peak TSH $f = 3.96$ $p < 0.05$.

Duncan's multiple range test (Table 2) showed a significant difference of basal and peak TSH levels between the two younger age groups as compared to the oldest age category ($p < 0.05$). The mean peak TSH response remained at the level of 36.4 mU/l until the third year, after which it declined to approximately normal levels.

GROWTH IN INFANTS AND CHILDREN WITH DOWN'S SYNDROME IN RELATION TO THYROTROPIN

The Baltimore study showed that growth was delayed in all DS subjects in relation to the standard population values. The deviation in growth is shown in

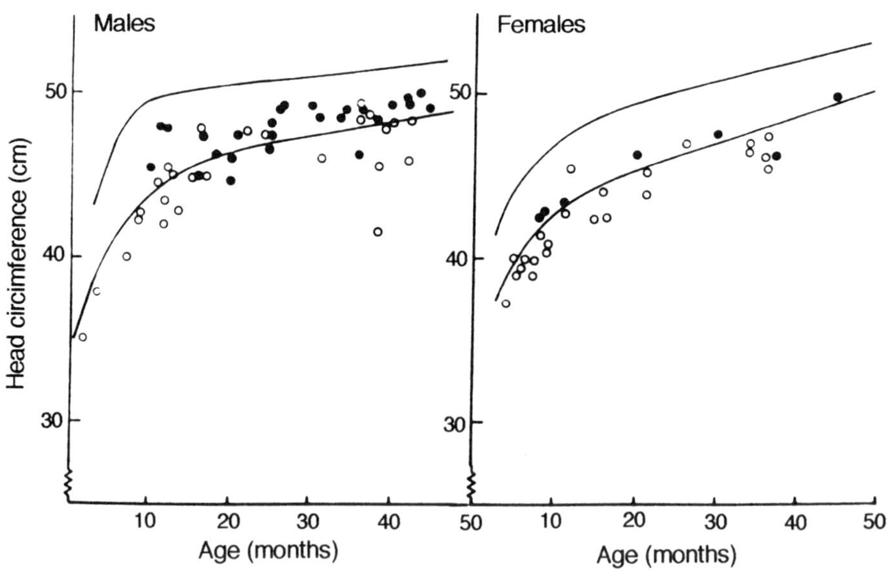

Figure 1. Head circumference in Down's syndrome infants. The head circumference of high TSH (open circles) and normal TSH infants (closed circles) is shown separately in male and female subjects. The range of standard values, i.e. 5th to 95th percentile, is indicated

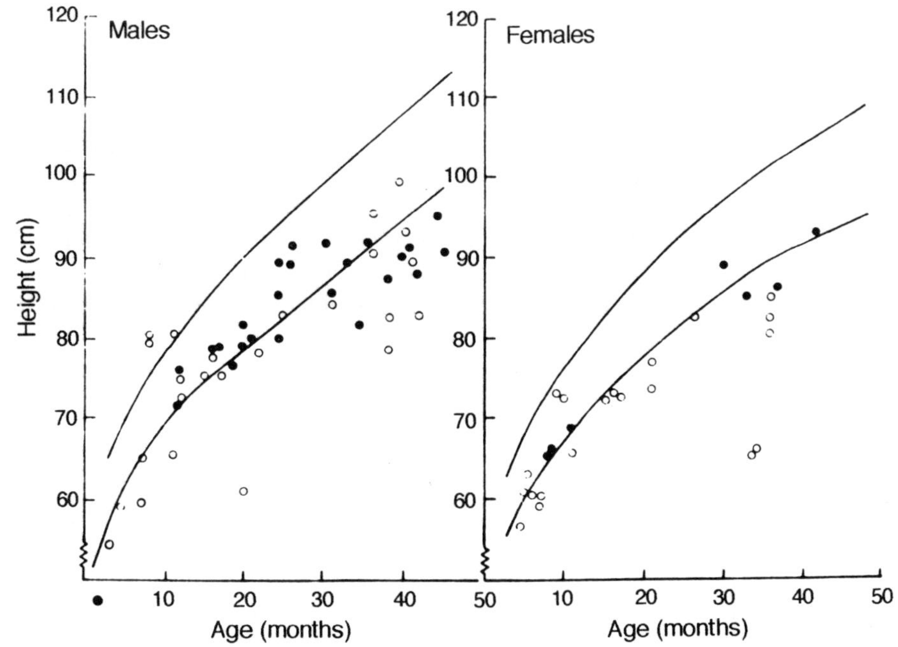

Figure 2. Height in Down's syndrome infants. The height of high TSH (open circles) and normal TSH infants (closed circles) is shown separately in male and female subjects. The range of standard values, i.e. 5th to 95th percentile, is indicated

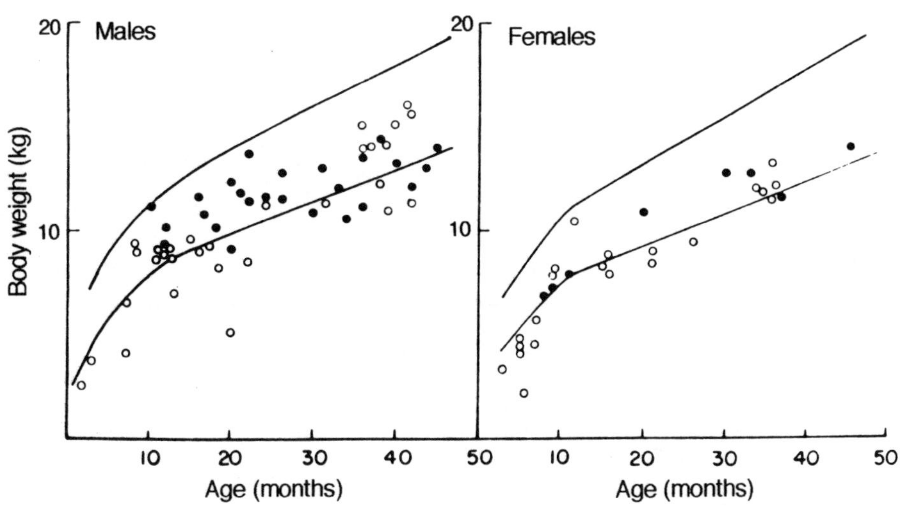

Figure 3. Weight in Down's syndrome infants. The weight of high TSH (open circles) and normal TSH infants (closed circles) is shown separately in male and female subjects. The range of standard values, i.e. 5th to 95th percentile, is indicated

Figs 1–3. The delay was more prominent in the HTSH group. When the deviation in growth of all parameters was expressed as a fraction of SD appropriate for age and sex it was found that the average decrement of the HTSH group was significantly greater than either normal children or NTSH DS ($p < 0.05$). In subjects in the four months to two years subgroup a significant correlation was observed between the elevated level of blood TSH and the degree of retardation in growth (Figs 4–6). The *t* values are indicated in individual figures. The level of blood TSH was related to all growth parameters. All growth parameters correlated inversely with TSH concentration. This correlation was not significant after two years of age and was absent among NTSH DS subjects.

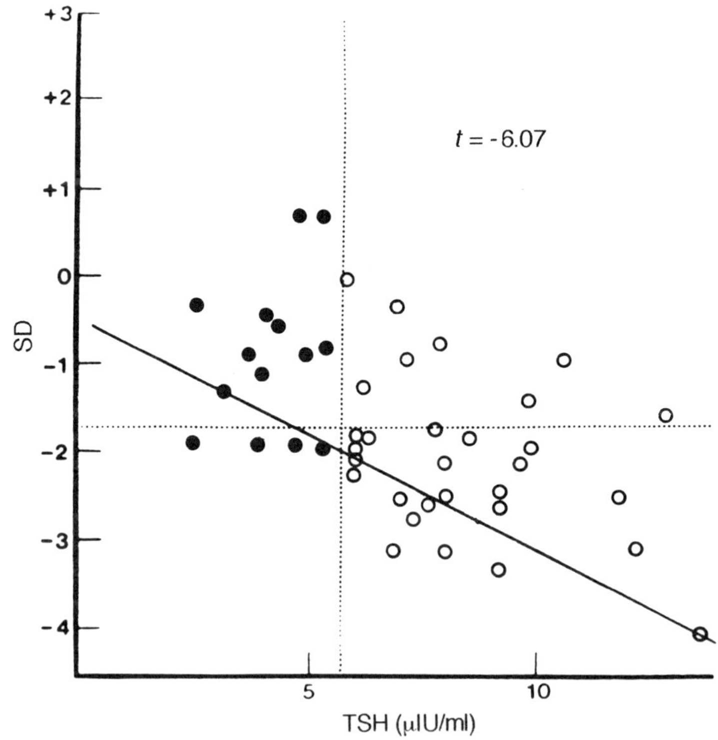

Figure 4. Relationship between deviant growth of head circumference and the blood TSH values in Down's syndrome infants between four months and two years. The TSH values are shown on the abscissa. In each subject the retarded growth is expressed as a fraction of the appropriate standard deviation for age and sex; the decrement (– SD) or increment (+ SD) is shown on the ordinate. The vertical line indicates the separation between high TSH (open circles) and normal TSH (closed circles), and the horizontal indicates the fifth percentile. The *t* value is indicated

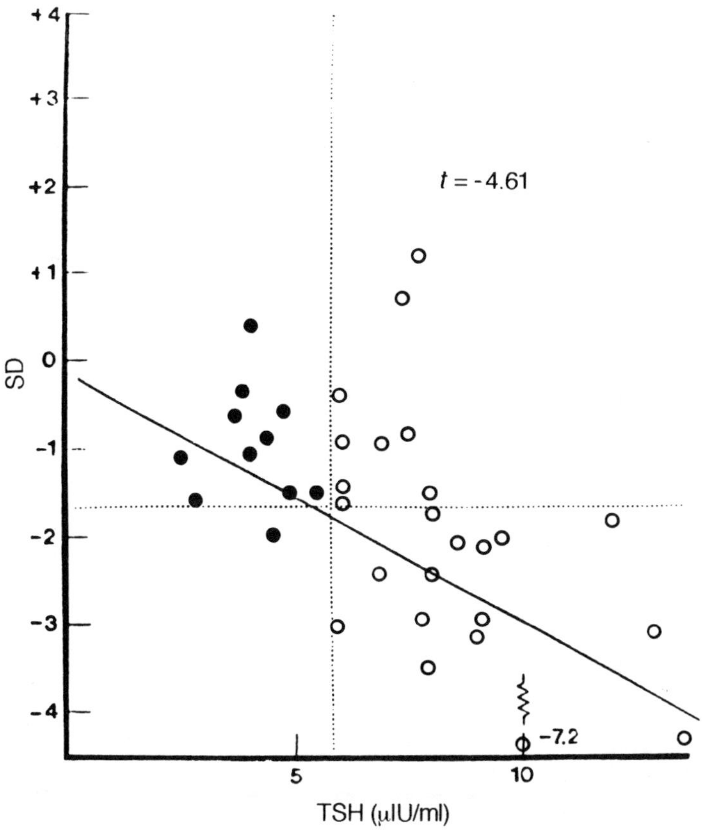

Figure 5. Relationship between deviant growth of height and the blood TSH values in Down's syndrome infants between four months and two years. Calculations are as in Fig. 4

EFFECT OF THYROID TREATMENT ON GROWTH IN DOWN'S SYNDROME WITH HIGH THYROTROPIN

Male Down's Syndrome

Of the 19 males 13 showed high basal TSH and an exaggerated response to TRH stimulation; of these latter subjects seven were given treatment with L-thyroxine and six had no treatment. Six DS subjects had normal basal TSH and a normal response to TRH stimulation. There were no differences in their T_4 measurements (Table 3). When growth was studied in these three groups of male DS subjects, i.e. HTSH treated and non-treated, and NTSH, no significant differences could be found in their growth parameters or rate of growth. The growth of our population was in the range of the DS growth charts and fifth percentile of the general population (18).

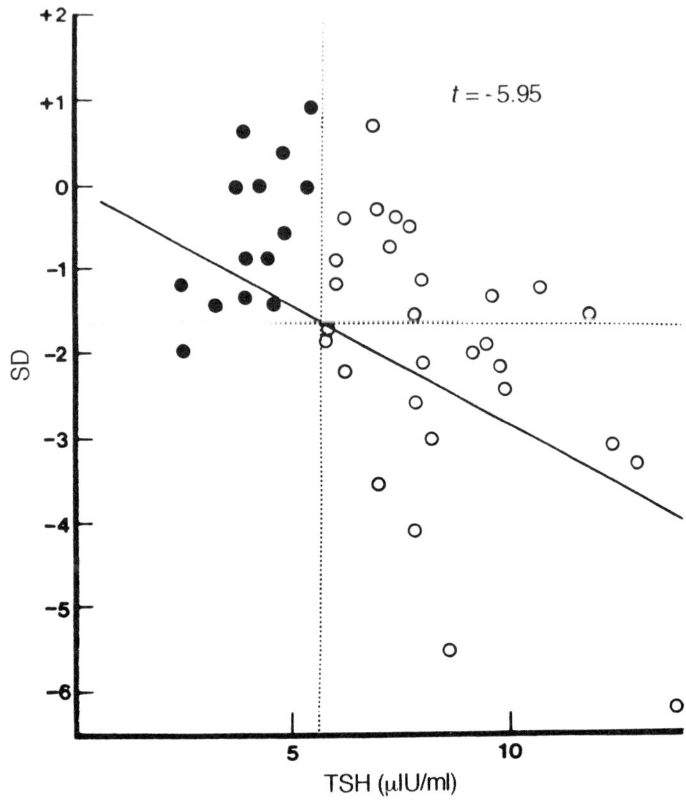

Figure 6. Relationship between deviant growth of weight and the blood TSH values in Down's syndrome infants between four months and two years. Calculations are as in Fig. 4

Table 3. T_4 basal TSH and peak TSH response in three groups of male DS subjects

	No.	T_4 (nmol/l)	Basal TSH (mU/l)	Peak TSH (mU/l)
High TSH Tx	7	149 ± 27	8.2 ± 2.7	39.3 ± 8.5
High TSH	6	141 ± 21	8.6 ± 1.7	31.6 ± 31.6
Normal TSH	6	147 ± 28	2.6 ± 0.6	22.2 ± 0.7

Table 4. T_4 basal TSH and peak TSH response in three groups of female DS subjects

	No.	T_4 (nmol/l)	Basal TSH (mU/l)	Peak TSH (mU/l)
High TSH Rx	5	$131 \pm 14^*$	$13.3 \pm 6.2^{**}$	42.6 ± 6.5
High TSH	5	144 ± 26	$6.9 \pm 2.0^{**}$	43.8 ± 6.9
Normal TSH	7	$184 \pm 36^*$	2.4 ± 1.2	18.4 ± 6.5

Student's *t*-test: $^*p < 0.03$; $^{**}p < 0.05$.

Female Down's Syndrome

Of the 17 females, 10 showed a high basal TSH and an exaggerated TSH response to TRH stimulation, and were divided into five who received treatment with L-thyroxine and five who were not given treatment. Seven had a normal basal TSH and a normal response to TRH stimulation and had no treatment. On analysis it was found that the basal TSH in the HTSH treatment group was significantly higher ($p < 0.05$) than in the non-treatment group. In addition, although T_4 in the treatment group was in the normal range for age, it was lower than the T_4 of the non-treatment group and was significantly lower ($p < 0.03$) than the NTSH group (Table 4). This suggested that the thyroid dysfunction was more severe in the HTSH treatment group than the non-treatment group.

Growth studies showed that at one month of age there were no differences in head circumference, height and weight between the three groups. However, rate of growth for head circumference ($p < 0.01$) and height ($p < 0.05$) during the first five months of life was significantly different in the treated group as compared to the non-treated group (Table 5). There was no significant difference in weight or rate of weight gain. The rate of growth of the HTSH treated

Table 5. Differences in rate of growth of head circumference and height in the first six months of life in three groups of female DS subjects

	High TSH Rx	High TSH	Normal TSH
Head circumference			
One month	34.7 ± 1.6	34.8 ± 0.5	34.5 ± 1.0
Six months	39.9 ± 1.3	37.9 ± 0.7	40.5 ± 1.9
Rate of growth	5.2^*	3.1^*	6.0
Height			
One month	51.4 ± 3.9	52.4 ± 1.5	50.7 ± 1.9
Six months	62.8 ± 1.4	61.0 ± 2.4	61.4 ± 3.3
Rate of growth	11.4^*	8.6^*	10.7

Student's *t* test: $^*p < 0.01$; $^{**}p < 0.05$.

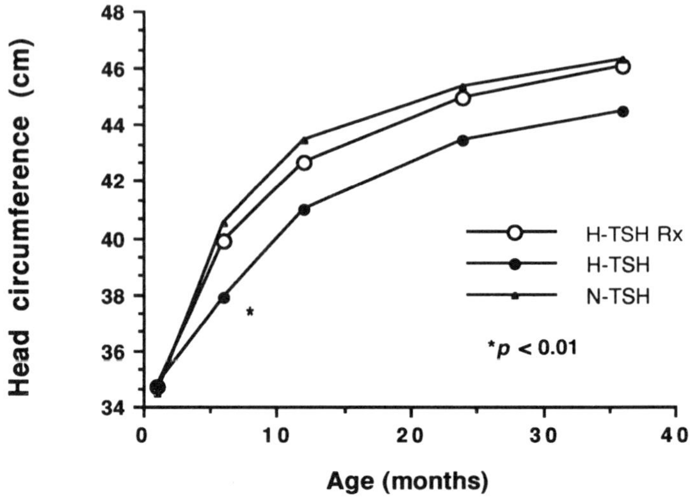

Figure 7. Growth of head circumference in three groups of Down's syndrome females from the age of one month to three years: high TSH Rx (receiving thyroid supplement), high TSH with no supplement, and normal TSH

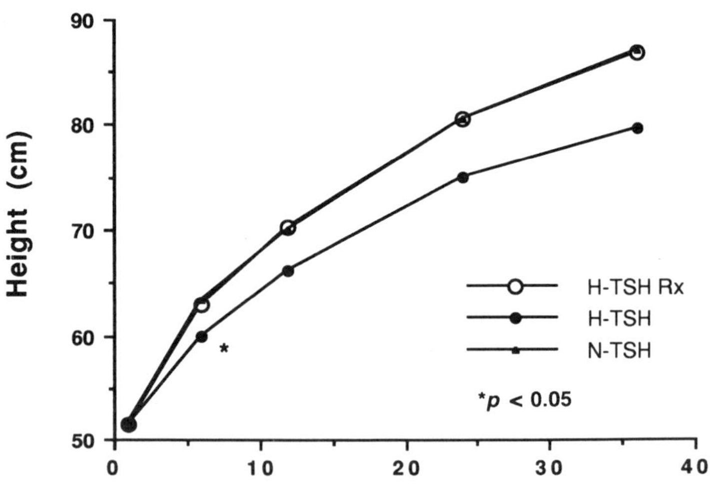

Figure 8. Growth of height in three groups of Down's syndrome females from the age of one month to three years: high TSH Rx (receiving thyroid supplement), high TSH with no supplement, and normal TSH

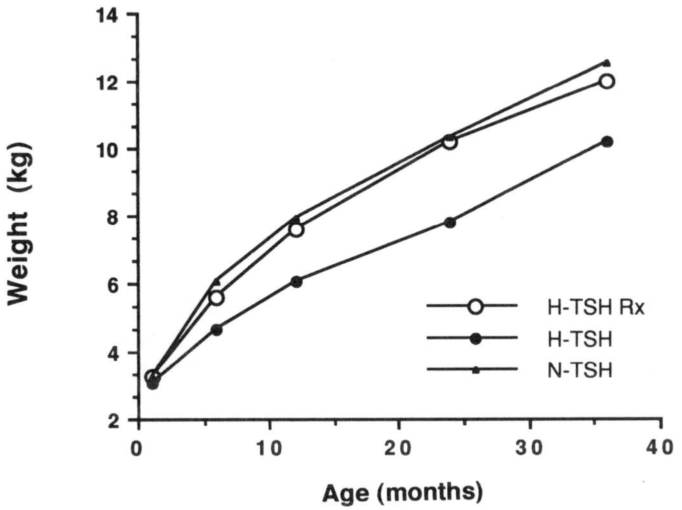

Figure 9. Growth of weight in three groups of Down's syndrome females from the age of one month to three years: high TSH Tx (receiving thyroid supplement), High TSH with no supplement, and normal TSH

group was similar to that of the NTSH group. There were no differences in rate of growth in any of the three groups following the first six months. The non-treated HTSH group showed a rate of weight gain less than the other two groups but it was not significant (Figs 7–9).

COMMENT

When discussing the relationship between thyroid dysfunction and DS in infancy one has to keep in mind the normal developmental sequence of the fetal hypothalamic–pituitary–thyroid system and the variability and changes in secretion pattern of thyroid hormones both *in utero*, at birth and postnatally. Whether DS subjects have the same pattern and sequence of development has yet to be shown. The DS thyroid gland has been described as being small in size and histologically poorly defined (1). Furthermore, DS subjects have retardation of brain growth, maturation delay and cortical dysgenesis (19). This could result in differences in the functioning and maturation of the fetal hypothalamic–pituitary–thyroid system in relation to the normal infant population. Thyroid dysfunction may just be one expression of this difference. The fetal hypothalamic–pituitary–thyroid system develops independently of maternal influence, with the placenta being impermeable to maternal hormones (13). Thyroid ontogenesis is divided into three phases. (a) Embryogenesis is completed by 12 weeks, with the thyroid gland having developed its histological features and ability to concentrate radioiodine and synthesize iodothyronines. At the same time the fetal pituitary contains TSH. (b) Hypothalamic maturation includes the development of the pituitary portal vascular system from 7 weeks to 30–35 weeks. TRH and somatomedin are measurable by 8–10 weeks and increase progressively in concentration to term. (c) Hypothalamic–pituitary–thyroid system control and the pituitary portal vascular system develop simultaneously from plexuses over the inferior hypothalamus and pituitary gland and by 30–35 weeks the system is interconnected.

During the period of maturation of negative feedback control T_4 increases in the serum from 20 to 40 weeks gestation and T_3 between 30 and 40 weeks Serum TSH levels peak at 20–24 weeks at about 15 mU/l and decrease to 10 mU/l at term.

Hence the human fetus matures from a state of combined primary hypothyroidism and hypothalamic–pituitary hypothyroidism during the first trimester to hypothalamic hypothyroidism during the last half of gestation to a state of mature function. The full-term fetus *in utero* manifests an exaggerated and prolonged TSH response when TRH is administered to the mother.

The biochemical study of thyroid function in DS subjects shows that they have T_4 and T_3 levels that are appropriate for age associated with persistently higher basal TSH levels than normal controls. In the Baltimore study HTSH was associated with significantly higher T_4 than in the NTSH group, although the T_4 values were in the age-appropriate range. This finding was not repeated in the Jerusalem studies of thyroid function, where HTSH was accompanied by normal or even somewhat low T_4 values. Although there does not seem to be a clear-cut explanation for these differences in T_4 and since all values were within the age-appropriate range it may just reflect the variability of thyroid hormone values in the young age group.

Our findings of an increased incidence of high TSH values or thyroid dysfunction in DS confirm those of a now increasing number of studies (8, 9, 20, 21). When we studied this dysfunction longitudinally there was a decline in basal TSH with time but at a much slower rate than in the general population. TSH peaks at birth and continues to be physiologically high in the newborn. Levels decline to normal levels by four days of age and even transiently high TSH levels return to normal by two months of age (22). In DS this process may be prolonged beyond two to three years. However, by the age of three years most DS infants who initially had elevated basal TSH values have values in the normal range. The higher basal TSH is accompanied by an exaggerated peak TSH response which again is higher than the peak response of an age-matched control population. The values of our control population are comparable to those of a larger population of the same age range (23). The peak TSH response to TRH stimulation continues to be exaggerated to a later age than the basal TSH, until it too declines to what are considered normal values. These values, though within the normal range, continue to be significantly more elevated than that of the normal population and may be indicative of continuing mild degrees of borderline thyroid dysfunction in DS (Table 1).

In the DS infant, prolongation of the physiologically high TSH could be evidence of immaturity of the hypothalamic–pituitary axis. This could result in the secretion of a TSH of low biological activity, which in turn may be an expression of a TRH deficiency (24, 25, 26). Other hypothalamic factors involved in TSH regulation, such as dopamine or somatostatin, may be involved. Delayed maturation of the hormone somatomedin has been shown to occur in DS (27) with incomplete switching from the fetal form of somatomedin to the growth hormone-related insulin growth factor. Transiently elevated TSH levels have been observed in prematures, associated with jaundice, and with respiratory and fetal distress (28). The cause of the thyroid insufficiency as yet remains unknown. The suggested hypothesis is that it is due to low thyroidal content of iodine in newborn infants and delayed maturation of enzymatic mechanisms of intrathyroidal hormonogenesis. Although none of the DS infants or children studied were premature or ill, the above could be considered as possible etiological factors, especially in view of the poorly defined histology of the DS thyroid gland. At a later age other compensatory mechanisms may come into play, such as reduction in the pituitary threshold for the feedback of TSH (29, 30).

Infancy is the most active period of growth, especially the first six months of life. The greatest growth in head circumference is completed by the age of two years. The underdeveloped and poorly histologically defined thyroid gland in DS may be more vulnerable and stressed during periods of active growth such as early infancy and perhaps later at puberty. This stress may be reflected in the biochemical profile of thyroid hormones, especially in the TSH values, which are the most sensitive indicator of subclinical hypothyroidism. As the accelerated growth rate of infancy declines so too the demand for thyroid hormones is less and the borderline functioning of the DS hypothalamic–pituitary–thyroid axis may be sufficient for the lesser growth demands. Thyroid dysfunction as reflected by TSH values improves. In the Baltimore study greater growth retardation was associated with higher TSH values, the HTSH group showing more retardation of growth than the NTSH group. The younger the infants the greater the correlation with high TSH values. Under the age of two years high TSH values were significantly correlated with growth retardation ($p < 0.05$) for all growth parameters. This correlation was no longer significant after two years of age. Females were more affected than males. Other factors affecting growth may begin to play a relatively more important role. Among these factors one has to consider growth hormone, familial or genetic influences nutrition and infections.

The preliminary analysis of the Jerusalem study is in keeping with the Baltimore findings although one hesitates to make firm conclusions from such small numbers. The males were again less affected than the females, showing a less severe picture of thyroid dysfunction. There were no significant differences in growth in the three groups of male DS subjects and thyroid supplementation had no measurable effect. Analysis of the data on the female DS subjects showed some interesting findings. All three groups of DS subjects had head circumferences of the same size at one month of age (Table 3). It is interesting to note that hypothyroidism, unlike DS, is not associated with a small head circumference (31). So even in hypothyroid DS one would not necessarily expect a small head circumference. The DS females with HTSH who were allocated to the treatment group were found to have a significantly lower T_4 ($p < 0.03$) than the DS females in the NTSH group, even though the T_4 levels were in the age-appropriate range. The HTSH treatment group also had a significantly higher basal TSH ($p < 0.05$) than the non-treatment group. Both these findings might suggest a more severe degree of thyroid decompensation in the HTSH group even prior to thyroid supplementation. It was in this group that the rate of growth of head circumference and height in the first six months of life was significantly greater than the non-treated group and approximated that of the NTSH group. After the first six months of life no differences in rate of growth could be shown between any of the three groups. The early gain in growth continued to be reflected in the growth percentiles. This may indicate that for thyroid supplementation to be effective it has to be given very early and treatment effects will only be seen in more severely affected individuals. One is still left with the dilemma of deciding which are the more severely affected individuals in light of the variability of thyroid hormone values during the period of infancy and the influence of other variables affecting growth. The question of giving thyroid supplementation in DS has been controversial since false hopes of a cure were raised. Treatment has been recommended for other groups of non-DS infants with a biochemical picture of thyroid dysfunction. Among these groups have been premature infants (28) and infants with hypopituitarism (32). Thyroid supplementation in a group of infants with pituitary insufficiency and with a biochemical picture of thyroid dysfunction led to an increased growth rate during the period of treatment. This group of infants had normal T_4 values associated with higher than normal basal TSH and an exaggerated response to TRH, a biochemical profile similar to that in DS. This may be further evidence of the contribution of the pituitary to thyroid dysfunction in the DS infant.

In conclusion it is suggested that both the thyroid gland and the development of the fetal pituitary–hypothalamic–thyroid system may differ in structure and maturation rate in the DS population, with thyroid dysfunction being one measurable expression. When the system is stressed by greater metabolic demands such as the active growth period of infancy the picture of thyroid dysfunction is more marked.

ACKNOWLEDGEMENTS

My grateful thanks to Prof. Alex Russell for his constant interest and encouragement and to the Balint foundation for its support.

REFERENCES

1. Benda CE, Clemens E. *The Child with Mongolism*, Grune & Stratton, New York, 1960.
2. Koch R, Share J, Graliker B. The effects of cytomel on young children with Down's syndrome (mongolism): A double blind longitudinal study. *J Pediatr* 1965; **66:** 776–778.
3. Murdoch JC, Ratcliffe WA, McLauty DG *et al.* Thyroid function in adults with Down's syndrome. *Biol Psychiatry* 1979; **14:** 463–471.
4. Baxter RG, Larkins RG, Martin FIR. Down's syndrome and thyroid function in adults. *Lancet* 1975; **ii:** 794–796.
5. Sare Z, Ruvalcaba RH, Kelly VC. Prevalence of thyroid disorders in Down's syndrome. *Clin Genet* 1978; **14:** 154–158.
6. Failkow PJ, Thuline HC, Hecht F *et al.* Familial predisposition to thyroid disease in Down's syndrome: controlled immunoclinical studies. *Am J Hum Genet* 1986; **23:** 67–86.
7. Fort P, Lifshitz R, Bellisario R *et al.* Abnormalities of thyroid function in infants with Down's syndrome. *J Pediatr* 1984; **104:** 545–549.
8. Pueschel SM, Pezzulo JC. Thyroid dysfunction in Down's syndrome. *Am J Dis Child*; **139:** 636–638.
9. Cutler T, Benezra-Obeiter R, Brink SJ. Thyroid function in young children with Down's syndrome. *Am J Dis Child* 1986; **140:** 479–483.
10. Sharav T, Landau H, Zadik Z *et al.* Age-related patterns of thyroid-stimulating hormone response to thyrotropin-releasing hormone stimulation in Down's syndrome. *Am J Dis Child* 1991; **145:** 172–175.
11. Tirosh E, Taub Y, Scher A *et al.* Short-term efficacy of thyroid supplementation for patients with Down's syndrome and low-borderline thyroid function. *Am J M Ret* 1989; **93:** 652–656.
12. Klein AH, Meltzer S, Kenny FM. Improved prognosis in congenital hypothyroidism treated before age three months. *J Pediatr* 1972; **81:** 912–915.
13. Fisher AD. Thyroid physiology and function tests in infancy and childhood. In *The Thyroid: A Fundamental and Clinical Text*, edited by Wermer SC, Ingabar SH, Harper & Row 1978; pp. 375–383.
14. Patel YC, Burger HG, Hudson B. Radioimmunoassay of serum thyrotropin: sensitivity and specificity. *J Clin Endocrinol Metab* 1971; **31:** 768–773.
15. *SAS/STAT Guide for Personal Computers*, Version 6, SAS Institute, Cary, NC, 1985; pp. 56–82.
16. Cole JWL, Grizzle JE. Application of multivariate analysis of variance of repeated measurements. *Biometrics* 1966; **122:** 810–828.
17. Sharav T, Collins RM, Baab PJ. Growth studies in infants and children with Down's syndrome and elevated levels of thyrotropin. *Am J Dis Child*, 1988; **142:** 1302–1306.
18. Cronk C, Crocker AC, Pueschel SM *et al.* Growth charts for children with Down's syndrome: 1 month to 18 years of age. *Pediatrics* 1988; **81:** 102–110.

19. Winiewski KE. Down's syndrome children often have brain with maturation delay, retardation of growth and cortical dysgenesis. *Am J Med Genet Suppl* 1989; **7:** 274–281.
20. Loudon MM, Day RE, Duke EMC. Thyroid dysfunction in Down's syndrome. *Arch Dis Child* 1985; **60:** 1149–1151.
21. Pozzan GB, Rigon ME, Girelli ME *et al.* Thyroid function in patients with Down's syndrome: preliminary results from non-institutionalized patients in the Veneto region. *Am J Med Genet Suppl* 1990; **7:** 57–58.
22. Barmasch M, Kuschiner A, Fontana O *et al.* Elevated neonatal thyrotropin (TSH) values in non hypothyroid newborns: differentiation with congenital hypothyroidism. Iodine deficiency disorders and congenital hypothyroidism. *Proceedings of the Satellite Meeting, 9th International Thyroid Congress*, Sao Paulo, Brazil, 1985; pp. 276–283.
23. Wolter R, Bourdoux R, Ermans AM. TSH response to TRH according to age. *Pediatr Res* 1982; **16:** 896.
24. Faglia G, Bitensky L, Pinchera A *et al.* Thyrotropin secretion in patients with central hypothyroidism: evidence for reduced activity in immunoreactive thyrotropin. *J Clin Endocrinol Metab* 1979; **48:** 989–996.
25. Illig R, Krawczynska H, Torresani T *et al.* Elevated plasma TSH and hypothyroidism in children with hypothalamic hypopituitarism. *J Clin Endocrinol Metab* 1977; **44:** 453–458.
26. Beck-Peccoz P, Amr S, Menezes-Ferreira M *et al.* Decreased receptor binding of biologically inactive thyrotropin in central hypothyroidism. *N Engl J Med* 1985; **312:** 1085–1090.
27. Sara V, Gustavson KH, Anneren G *et al.* Somatomedins in Down's syndrome. *Biol Psychiatry* 1983; **18:** 803–811.
28. Delange F, Dodion J, Walter R *et al.* Transient hypothyroidism in the newborn infant. *J Pediatr* 1978; **92:** 974–976.
29. Sato T, Sluzuki J, Takentani T *et al.* Age related change in pituitary threshold for TSH release during thyroxine replacement therapy for cretinism. *J Clin Endocrinol Metab* 1977; **44:** 553–559.
30. Schultz RM, Glassman MS, MacGillivary MH. Elevated threshold for thyrotropin suppression in congenital hypothyroidism *Am J Dis Child* 1980; **134:** 19–20.
31. Letarte J, Guyda H, Dussault JH. Clinical, biochemical and radiological features of neonatal hypothyroid infants. In *Neonatal Thyroid Screening*, edited by Burrow GN, Raven Press, New York, 1980; pp. 25–49.
32. Gruneiro de Papendieck L, Iorcansky S, Rivarols MA *et al.* Patterns of TSH response to TRH in children with hypopituitarism. *J Pediatr* 1982; **100:** 387–392.

Discussion of Presentation

THERESA SHARAV
(Edited by D.A. Snider and Salvador Castells)

R. McVie: Among children with congenital hypothyroidism that we pick up through screening programs, only about eight in 200 cases can be identified clinically. Their head circumferences are large, so there is a fascinating difference with the Down's syndrome children where the TSH or TRH stimulation is abnormal.

T. Sharav: The hypothyroid infants have been shown to have head circumferences well up into the 75th percentile. But if we consider that Down's syndrome is supposed to have intrauterine growth retardation, then at a later age they become microcephalic relative to the head circumference at birth, which is within the fifth percentile of the normal population.

N. Stahnke: Interpretation of your data is quite difficult. You found normal T_4 levels and elevated TSH values; maybe there was some kind of T_4 insensitivity at this central region or at the periphery. The more important hormone is the T_3. Do you have data on the T_3 levels in these patients and do you have data on the free hormone concentrations?

T. Sharav: In the original studies, we looked at the serum T_3 concentrations and we found no difference between normals and Down's syndrome children. There are now several papers that describe this phenomenon of thyroid dysfunction with high serum TSH levels. There are various papers which confirm that serum T_3 and the T_4 levels are within the normal range. We have not studied the cellular responses to thyroid hormones. I think this is very interesting, and it was one thing we did consider—whether there was a peripheral resistance to thyroid hormone in Down's syndrome because we know that the cell membrane in Down's syndrome is different from the normal child's membrane. We thought of receptor difficulties, but I think the problem reflects a central mechanism more than a peripheral mechanism.

7 Thyroid and Immune Functions in Persons of Different Ages with Down's Syndrome

MÁRIA ŠUSTROVÁ AND VLADIMÍR ŠTRBÁK[1]

Department of Clinical Immunology, Children's Hospital, and [1]*Institute of Experimental Endocrinology, Slovak Academy of Sciences Bratislava, Slovakia*

INTRODUCTION

Down's syndrome (DS), the phenotypic expression of human trisomy 21, is manifested by morphogenetic abnormalities by immunodeficiency, by mental retardation and the pathological manifestations of Alzheimer's disease (3–8). DS children are at increased risk for leukemia and have a greater incidence of repeated and chronic respiratory infections and autoimmune disorders than the general population (1, 10).

The goal of our work was to detect early signals of changes in immune and thyroid (2, 9, 11) functions. We studied the changes in humoral and cellular immunity and the relationship between hypothyroidism and immunological defects associated with low serum levels of zinc (1, 7).

PATIENTS AND METHODS

We studied 110 subjects with DS (ages from 1 to 35 years) compared with an age-matched control group. All children and adults studied had lived at home all their lives, and were negative for the hepatitis B surface antigen. The patients were divided into three groups: DS1 (age one to six years), DS2 (age 6–15 years) and DS3 (over 15 years).

RESULTS

Increased levels of plasma triiodothyronine (T_3) in DS1 and decreased levels in DS3 were found (Table 1). Low levels of thyroxine (T_4) were found in DS3 (Table 2). Increased plasma thyrotropin (TSH) and thyroxine-binding globulin (TBG) were found in all age groups (Tables 3 and 4). Hypothyroidism (clinical

Growth Hormone Treatment in Down's Syndrome. Edited by S. Castells and K.E. Wisniewski

and biochemical) was found in 16 children and 10 adults (23.6%). Increased plasma immunoglobulin A (IgA) levels were found in DS groups (261%, 180% and 173.3% in DS1, DS2 and DS3 respectively, $p < 0.001$). In contrast, average plasma zinc levels were lower in DS (74.4%, 77.7% and 88.5% in DS1, DS2 and DS3 respectively, $p < 0.001$). We also found a significant association between low serum levels of zinc and hypothyroidism.

Table 1. T_3 levels in DS subjects and age-matched controls (mg/ml)

Group	n	$\bar{x}$	SD	p
DS1	25	1.85	0.57	<0.05
Controls	19	1.664	0.33	
DS2	38	1.71	0.64	n.s.
Controls	40	1.56	0.33	
DS3	36	1.04	0.49	<0.05
Controls	40	1.39	0.48	

Table 2. T_4 levels in DS subjects and age-matched controls (μg/dl)

Group	n	$\bar{x}$	SD	p
DS1	25	11.40	6.01	n.s.
Controls	19	10.7	3.59	
DS2	38	8.66	5.33	n.s.
Controls	40	8.34	1.76	
DS3	36	5.88	2.21	<0.001
Controls	40	8.23	1.76	

Table 3. TSH levels in DS subjects and age-matched controls (μU/ml)

Group	n	$\bar{x}$	SD	p
DS1	25	5.05	3.64	<0.001
Controls	19	2.08	1.09	
DS2	38	5.44	3.059	<0.02
Controls	40	3.058	2.44	
DS3	36	5.66	4.39	<0.001
Controls	40	1.39	0.49	

Table 4. TBG levels in DS patients and age-matched controls (μg/dl)

Group	n	$\bar{x}$	SD	p
DS1	25	17.7	6.2	<0.05
Controls	19	14.7	4.79	
DS2	38	16.5	4.58	<0.001
Controls	40	9.98	4.43	
DS3	36	11.56	5.53	<0.01
Controls	40	8.24	4.56	

DISCUSSION AND CONCLUSION

A positive correlation between TBG α-globulin concentration and high levels of IgA was found in children with DS. A disturbance in the regulation of IgA production (high plasma levels) might be reflected in the frequent finding of high serum TBG, T_4 and T_3 levels in the youngest patients with DS. A very low concentration of plasma zinc is usually connected with the low plasma T_3, sometimes without adequate TSH increase. Treatment with zinc was able to re-establish adequate TSH levels and to increase plasma T_3 only in some patients. We came to the conclusion that the changes in immune and thyroid functions are related to the age of persons with DS. All patients with DS deserve careful endocrinological and immunological follow-up to prevent secondary consequences of the disorders.

ACKNOWLEDGEMENTS

The authors wish to thank the National Hormone and Pituitary Program of the NIADDK for supply of Human TSH reagents. The project has been partially supported by SAS grant #999431.

REFERENCES

1. Björkstén B, Bäck O, Gustavson, KH, Hallman G, Hägglof B, Tärnvik A. Zinc and immune function in Down's syndrome. *Acta Paediatr Scand* 1980; **69:** 183–187.
2. Burgio GR, Severia F, Rossoni R, Vaccaro R. Mongolism and thyroid autoimmunity. *Lancet* 1965; **i:** 166.
3. Burgio GR, Ugazio AG, Nespoli I, Maccario AF. Down's syndrome: a model of immunodeficiency. *Birth Defects* 1983; **12:** 325–327.
4. Fabris N, Mocchegiani E, Amadio L. Thymic hormone deficiency in normal ageing and Down's syndrome: is there a primary failure of the thymus? *Lancet* 1984; **i:** 939–986.

5. Larocca LM, Piantelli M, Valitutti S, Castellino F, Maggiomo N, Musianai P. Alterations on thymocyte subpopulations in Down's syndrome (trisomy 21). *Clin Immunol Immunopathol* 1988; **49:** 175–176.
6. Montagna D, Maccario R, Ugazio AG, Nespoli L, Pedroni E, Fagiano P, Burgio G. Cell mediated cytotoxicity in Down's syndrome: impairment of allogeneic mixed lymphocyte reaction. NK and NK-like activities. *Eur J Pediatr* 1988; **148:** 53–57.
7. Noble RL, Warren RP. Analysis of blood cell populations plasma zinc and natural killer cell activity in young children with Down's syndrome. *J Ment Defic Res* 1988; **32:** 192–201.
8. Spangelo BL, McLeod R. Is the thymus an endocrine gland? *Prog Neuroendocrine Immunol* 1988; **1:** 16–20.
9. Ugazio AG, Jaykar S, Marcioni AF, Duse M, Monafo V, Pasquale F, Burgio GR. Immunodeficiency in Down's syndrome: relationship betwen presence of human thyroglobulin antibodies and HBsAg carrier status. *Eur J Paediatr* 1977; **126:** 139–149.
10. Ugazio AG, Maccario R, Norrarangelo LD, Burgio GR. Immunology of Down's syndrome: a review. *Am J Med Genet Suppl* 1990; **7:** 204–212.
11. Zori RT, Desmond AS, Oster H, Williams CA, Spillar R, Riley WJ. Relationship of autoimmunity to thyroid dysfunction in children with Down's syndrome. *Am J Med Suppl* 1990; **7:** 238–241.

8 Thyroid Function in Children with Down's Syndrome

SIEGFRIED ZABRANSKY
University of Saarland Medical Center, Division of Pediatric Endocrinology, Homburg/Saar, Germany

INTRODUCTION

Children with Down's syndrome (DS) show symptoms and signs which are usually to be seen in hypothyroidism (e.g. constipation, macroglossia, hypotonia, dry skin, weakness, short stature), suggesting a relationship between these two entities. Therefore, most doctors treat all children with DS with thyroxine, whether they have normal or abnormal serum values for thyroid hormones and thyrotropin (TSH). Published data about thyroid function in these patients are extremely controversial. During the last 10 years, therefore, we critically analyzed thyroid function of all patients with DS.

METHODS

The data of thyroid function were analyzed in 100 children (age 5 days to 16 years) with DS. The following parameters were determined: triiodothyronine (T_3), thyroxine (T_4), free thyroxine (FT_4) TSH before and after stimulation using intravenous (i.v.) application of 1 μg TSH-releasing hormone (TRH) per kilogram body weight (Radioimmunoassay, Amersham). In the TRH test, blood for evaluation of TSH was drawn before and 30 min and 60 min after

Table 1. Children with isolated elevated TRH-induced TSH secretion ($n = 15/100$; age 3–15 years)

Time after TRH injection (min)	TSH (U/L)
0	0.3–2.3
30	29–50
60	17–35

Growth Hormone Treatment in Down's Syndrome. Edited by S. Castells and K.E. Wisniewski

Table 2. Normal range for TSH and TRH-induced TSH

Age	TSH basal (μU/ml)	TSH 30 min after i.v. application of TRH/kg body weight (μU/ml)
2 weeks to 2 years	0.5–7	2–36
>2 years	0.5–7	2–20

In children younger than two years of age TRH-induced TSH secretion is higher than in older children (3).

Table 3. Results: children with normal values ($n = 82/100$)

Age	n		T_3 (ng/ml)	T_4 (μg/dl)	FT_4 (ng/dl)	TSH (μU/ml)	TSH 30 min (U/ml)	TSH 60 min (μU/ml)
5 days	9	($\bar{x}$)	4.1	12.6	3.0	8.8		
		(SD)	0.5	1.2	0.8	1.2		
3 months	12	($\bar{x}$)	1.8	10.7	0.8	5.0	20.8	14.6
		(SD)	0.3	1.3	0.4	0.9	5.1	7.4
2nd year	5	($\bar{x}$)	1.5	10.8	1.7	2.1	15.4	6.3
		(SD)	0.3	1.2	0.2	0.5	3.1	1.2
3–16 years	56	($\bar{x}$)	1.6	9.5	1.8	1.5	11.4	6.0
		(SD)	0.3	1.3	0.5	0.6	2.7	1.5

$\bar{x}$ = mean; SD = standard deviation.

Table 4. Children with isolated elevated TRH-induced TSH-secretion ($n = 15/100$)

Age (years)	n	Sex	T_3	T_4	FT_4	TSH	TSH 30 min	TSH 60 min
3	2	F	2.4	14.3	2.0	2.3	42	35
		M	2.0	12.1	1.8	1.3	50	28
5	3	F	1.8	10.2	1.6	0.8	35	23
		F	1.3	9.1	1.9	0.3	32	21
		M	1.2	10.1	1.1	1.4	45	32
6	2	M	1.5	8.5	1.4	2.3	30	18
		M	1.1	8.0	1.2	1.1	32	27
8	1	F	1.5	9.2	1.9	0.8	48	30
10	2	F	1.9	7.4	1.2	2.1	51	32
		F	1.0	7.3	1.8	0.9	39	19
12	3	F	2.0	9.1	1.3	0.5	29	20
		F	1.3	7.9	1.5	1.5	49	27
		M	1.1	5.6	1.6	1.3	28	17
13	1	F	1.7	7.3	1.8	0.9	34	23
15	1	F	1.4	6.1	1.5	0.5	33	24

Table 5. Clinical signs: children with goiters ($n = 7/100$)

5 with euthyroid function
2 with hypothyroid function
T_3, 0.8 and 0.5 ng/ml
T_4, 3.4 and 2.8 μg/dl
TSH, 56 and 67 μU/ml

None of the children examined had clinical signs or laboratory data indicative of Grave's disease.

an i.v. injection of 1 μg TRH (TRH-Relefact, Hoechst) per kilogram body weight.

DISCUSSION

In contrast to other authors, we cannot confirm a high incidence of hypothyroidism in children with DS. Only one newborn with DS had athyreosis diagnosed by screening tests. Two children showed hypothyroid goiters (pathogenesis is not clearly demonstrated, and may be caused by iodine deficiency). Using newborn TSH screening in Germany we found congenital hypothyroidism (hypoplasia or aplasia of the thyroid gland, and ectopic thyroid tissue) with an incidence of 1 : 2600. The exact proportion of newborns with DS and hypothyroidism is not known. With the exception of one report only (2), a higher rate of incidence has not been observed. In 15 of 100 children with DS we found normal T_3, T_4, FT_4 and TSH values. TRH-induced TSH values were elevated, which was an unexpected result. Usually, isolated elevated TRH-induced, TSH secretion means latent hypothyroidism. However, in DS there must be additional unknown reasons to explain these findings. Otherwise we would see more patients with manifest hypothyroidism.

CONCLUSION

The diagnosis of DS is not an indication for prophylactic medication of T_4. We treat only verified cases of hypothyroidism with T_4. Isolated TRH-induced TSH secretion should not be the indication for T_4 treatment. Note that healthy children younger than two years of age show higher TRH-induced TSH secretion, and therefore the normal ranges have to be adapted to the age of the child. However, it is necessary that patients with DS should be examined for thyroid function.

REFERENCES

1. Aldenhoff P, Waldemaier C, Zabransky S, Helge H. Der TRH-Stimulationstest bei Kindern mit Kindern und Erwachsenen. *Mschr Kinderheilk* 1977; **125:** 544–545.
2. Fort P, Lifshitz F, Bellisario R, Davis J, Lanes R, Pugliese M, Richman R, Post EM, David R. Abnormalities of thyroid function in infants with Down's syndrome. *J Pediatri* 1984; **107:** 545–549.
3. Zabransky S, Lengle D. Die optimale TRH-Testdosis beträgt bei Kindern 1 μg/kg (i.v.); *Extracta Paediatr (Berlin)* 1980; **4(6):** 321–325.

9 Abnormal Thyrotropin Secretion in Down's Syndrome

A. CRINÒ,* P. CIAMPALINI,[2] M. C. DiGILIO,[1]
M. G. OBREGON, A. GIANNOTTI[1] AND P. BORRELLI
Reparto di Endocrinologia, Servizio di Genetica Medica[1], Ospedale 'Bambino Gesù'—IRCCS, Endocrinologia 1[2], Università 'La Sapienza', Rome, Italy

INTRODUCTION

An increased incidence of thyroid diseases has been observed in patients with Down's syndrome (DS) (hypothyroidism, Hashimoto's thyroiditis, Graves' diseases etc.), often depending on autoimmune pathologies or anatomical glandular alterations (hypoplasia, etc.) (1–3).

Particularly in children with this syndrome, congenital hypothyroidism is about 28 times more frequent (1 : 141) than in the general population (1 : 3800)(4); on the other hand, hyperthyroidism is a rare condition (only about 35 cases have been reported) (5–7).

High levels of basal thyrotropin (TSH) have recently been found in patients with DS, while triiodothyronine (T_3), thyroxine (T_4), free triiodothyronine (FT_3) and free thyroxine (FT_4) are within the normal range and thyroid autoantibodies (TAA) are negative (8–10).

The etiopathogenesis of such biochemical abnormalities, certainly different from cases with overt or subclinical hypothyroidism, is presently unknown; nor is it established if these patients should be treated with substitutional therapy (8, 9).

The aim of our study was to check the alterations of thyroid function present in patients with DS, to define a possible etiopathogenesis for this hormonal abnormality and to establish which patients should be treated with substitutional therapy.

SUBJECTS AND METHODS

We investigated thyroid function in 199 home-reared children and adolescents with DS: 113 males and 86 females, aged between 3 months and 16.5 years

*Address for correspondence: Reparto di Endocrinologia, Ospedale Pediatrico 'Bambino Gesù', IRCCS, Piazza S. Onofrio 4, 00165 Rome, Italy.

Growth Hormone Treatment in Down's Syndrome. Edited by S. Castells and K.E. Wisniewski

(mean age ± SD: 5.8 ± 4.5 years). Pubertal stage (Tanner): prepuberal $n = 156$; puberal $n = 43$. The control group consisted of 60 children—35 males and 25 females—with constitutionally short stature, of the same age.

In each subject we measured T_3, T_4, FT_3, FT_4, TAA (thyroglobulin and microsomal), basal TSH and 30, 60, 90 and 120 min after TRH stimulus (200 μg i.v.). A TRH test was performed in 103 patients with DS, clinically euthyroid. TSH values of $\geqslant 5$ μU/ml, and/or a TSH response to the TRH test of $\geqslant 28$ μU/ml, were considered pathological.

We also evaluated some of the growth parameters using Tanner's standards and growth charts for children with DS edited by Cronk *et al.* (11). Height was expressed as standard deviation score (SDS) for chronological age. Bone age was calculated by the method of Greulich and Pyle (12). The dosage of thyroid hormones was ascertained using the radioimmunoassay (RIA) method, and TSH and thyroid autoantibodies with the immunoradiometric assay (IRMA) method, using commercial kits.

Informed consent to the study was obtained from the patients and their parents. Study protocol was approved by the Ethical Committee of Ospedale Pediatrico Bambino Gesù, IRCCS. Statistical analysis was performed using Student's *t*-test, χ^2 test, correlations and linear regression. All data are expressed as mean ± standard deviation (SD).

RESULTS

Fourteen children with DS (7%) had overt thyroid diseases (Table 1). The other 185 patients (93%), all clinically euthyroid, had peripheral thyroid hormones within the normal range.

These subjects were subdivided into two groups according to normal or pathological values of TSH (basal and/or after TRH stimulus) (Table 2):

1. Eighty-two patients (group A: 44%) showed significantly ($p < 0.0001$) higher TSH levels, basal and/or after TRH, compared with the control group; in these patients basal TSH values, when >5 μU/ml, occurred almost always between 5 and 10 μU/ml, and never reached levels seen in clinical hypothyroidism (Fig. 1). For this reason they were initially diagnosed to be affected by subclinical hypothyroidism.
2. The other 103 children (group B: 56%) revealed normal thyroid function.

During the TRH test, performed in 103/185 children, 60 patients with high basal TSH levels (group A) showed an exaggerated and prolonged response (Table 3).

In 43 patients with normal basal TSH levels (group B) the peak response to the TRH test (30 min) was not different from the control group, even if TSH values at 90 and 120 min were higher than those of the control group (prolonged and delayed response) (Table 3). This pattern of response to TRH

Table 1. Overt thyroid diseases in 14/199 children (7%) (seven males, seven females) with DS[a]

Diagnosis	Sex	Age (years)	Thyroid function	TAA (HTG-TMS)	Therapy
Grave's disease	F	9.6	Hyperthyroidism	+ +	PTU + surgery
Grave's disease	F	10	Hyperthyroidism	+ +	MTZ
Grave's disease	F	6	Hyperthyroidism	+	PTU
Absence of right lobe of thyroid	M	14.5	Hypothyroidism	–	L-T_4
Thyroid hypoplasia	M	1	Hypothyroidism	–	L-T_4
Ectopic thyroid gland	M	0.3	Hypothyroidism	–	L-T_4
Thyroiditis	F	12	Subclinical hypothyroidism	+ +	L-T_4
Thyroiditis	F	11.8	Subclinical hypothyroidism	+ +	L-T_4
Thyroiditis	F	6	Subclinical hypothyroidism	+ +	L-T_4
Thyroiditis	M	7.8	Subclinical hypothyroidism	+ +	L-T_4
Thyroiditis	F	15	Hypothyroidism	+ +	L-T_4
Thyroiditis	M	16	Hypothyroidism	+ +	L-T_4
Thyroiditis	M	6.2	Hypothyroidism	+ +	L-T_4
Thyroiditis + IDDM	M	8.5	Hypothyroidism	+ +	L-T_4 + insulin

[a]Abbreviations: HTG-TMS, anti-thyroglobulin and anti-microsomal autoantibodies; PTU, propiltiouracil; MTZ, metimazole; IDDM, insulin-dependent diabetes mellitus; L-T_4. L-thyroxine

Table 2. Clinical and auxological data on 185 children with DS at first observation

($n = 185$)	Sex (M/F)	Age (years)	Heitht (SDS for chronological age)	Delayed bone age
Group A ($n = 82$) High TSH	47/35	5.7 ± 4.8	−2.1 ± 1.1	30/78 (38%)
Group B ($n = 103$) Normal TSH	59/44	5.8 ± 4.3	−1.7 ± 1.0	26/92 (28%)
$M \pm$ SD		n.s.	$p < 0.05$	n.s.

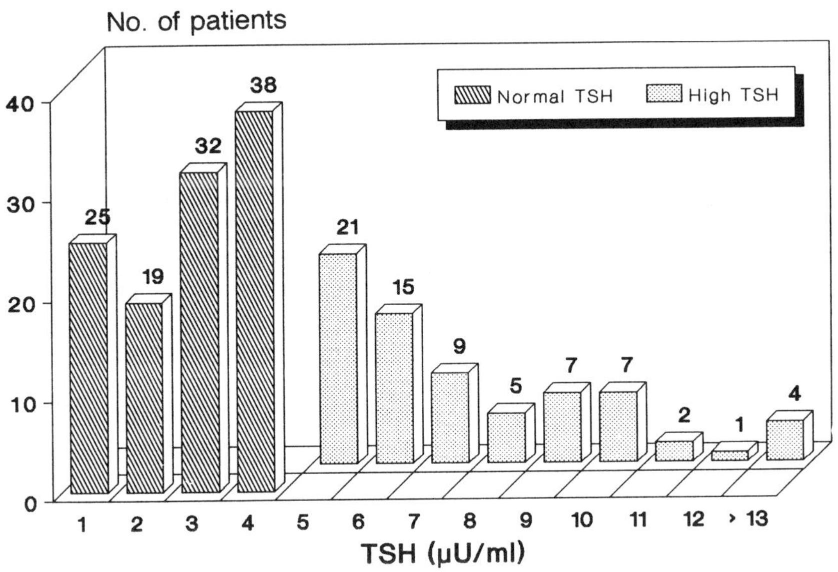

Figure 1. Basal TSH in 185 children with DS

is observed in several patients affected by hypothalamic hypothyroidism. We did not find a statistically significant link between age and high or normal TSH values.

Thyroid autoantibodies were present in 32/185 patients with DS (17%): in 18 patients with high TSH levels and in 14 patients with normal TSH values.

Thyroid function, with or without TRH test, was studied more than once in 76 children of group A (high TSH levels) and also in 81 of group B (normal TSH levels). Thyroid function reverted to normal in 35 children of group A (46%), without therapy, but in 24 of these TSH levels were fluctuating on

Table 3. TRH test in 103 children with DS, clinically euthyroid

	TSH (μU/ml) response to TRH (min)					FT_3 (pg/ml)	FT_4 (pg/ml)
	0	30	60	90	120		
Group A ($n = 60$) High TSH	$7 \pm 2.8^{**}$	$35.2 \pm 9.3^{**}$	$27.2 \pm 9.4^{**}$	$22.4 \pm 9.6^{**}$	$16.9 \pm 8.6^{**}$	3.6 ± 0.9	13.8 ± 3.2
Group B ($n = 43$) Normal TSH	3.04 ± 1.1	18.4 ± 4.8	14.1 ± 4.3	$10.6 \pm 3.5^{*}$	$7.9 \pm 2.8^{*}$	3.7 ± 1.0	14.1 ± 3.4
Controls ($n = 60$)	2.7 ± 1.4	17.0 ± 4.8	12.5 ± 3.9	8.6 ± 3.5	6.3 ± 2.4	3.8 ± 0.8	14.1 ± 2.8

$M \pm$ SD; $^{*}p < 0.01$; $^{**}p < 0.001$ versus controls.

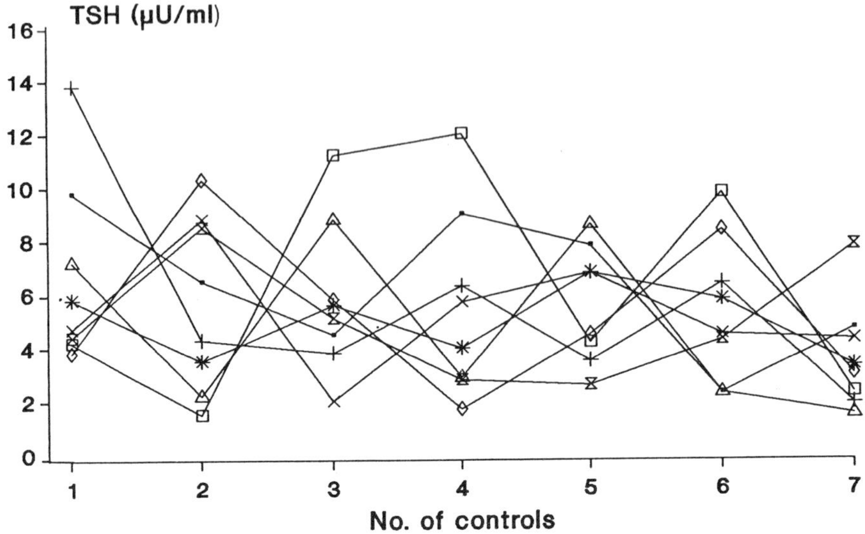

Figure 2. Eight subjects with DS: fluctuating TSH in the follow-up

follow-up. In fact, in some cases phases with high basal TSH levels were followed by periods of spontaneous normalization with TSH levels never higher than 12–14 μU/ml (Fig. 2). Nobody with persistently high TSH levels developed clinical hypothyroidism. Moreover, we found fluctuating basal TSH levels also in 25 of 81 patients in group B (normal thyroid function), in which it was possible to evaluate thyroid function on follow-up.

There was a statistically significant difference ($p < 0.05$) in height (SDS) between children with high TSH ($n = 82$) and those with normal TSH ($n = 103$), at first observation (Table 2). Only in 63 children of group A and in 73 of group B we could carry out a clinical and auxological follow-up: they

Table 4. Growth data on children with DS (follow-up) ($n = 136$)

	Age (years)		Height (SDS)		Decreased growth velocity	Fluctuating TSH
	At referral	At present	At referral	At present		
High TSH ($n = 63$) 39 M, 24 F	4.9 ± 4.2	7.3 ± 4.8	−1.9 ± 1.0	−2.0 ± 1.0	1 19/63 (30%)	20/63 (32%)
Normal TSH ($n = 73$) 48 M, 25 F	5.6 ± 4.2	7.4 ± 4.6	−1.8 ± 1.0	−1.8 ± 1.0	21/73 (29%)	23/73 (31%)
M ± SD	n.s.	n.s.	n.s.	n.s.	n.s.	n.s.

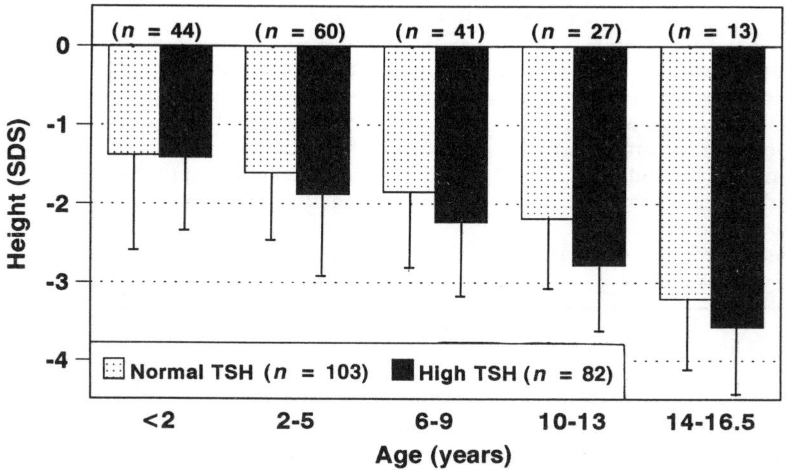

Figure 3. Height (SDS) by age groups in 185 children with DS

did not show any significant difference in height (SDS), either at referral or at the present time (Table 4).

Comparing height (SDS) and age groups, we observed a negative correlation, without significant differences between groups with normal or high TSH levels (Fig. 3). Also no significant difference was found between growth velocity and bone age in the two groups of patients (Tables 2 and 4).

DISCUSSION AND CONCLUSIONS

Our study shows a relatively high prevalence of overt thyroid disease in children and adolescents with DS (14/199 = 7%). On the other hand high basal TSH levels with FT_3 and FT_4 within the normal range, and without any clinical manifestation of thyroid diseases, represent a characteristic in some of these patients. In fact, we observed this hormonal abnormality in 44% of our population (82/185), in which TSH levels were almost always between 5 and 10 μU/ml and never higher than 14 μU/ml. This abnormal TSH secretion can sometimes be transitory (46%) and, particularly in some, can fluctuate with time.

It is interesting to remark also that some patients with DS, considered 'normal' from the point of view of thyroid function, could sometimes present with fluctuating TSH levels on follow-up. Such cases, without evidence of thyroid autoimmune diseases or anatomical glandular alterations, could be hypothesized as being due to a neuroregulatory disorder of TSH secretion, probably linked to altered availability of some cerebral neurotransmitters

Table 5. Possible etiopathogenesis of high TSH levels in DS

Thyroid disorders	Central disorders
Anatomical glandular alteration (hypoplasia, ectopy, etc.)	Neuroregulatory disorders of TSH secretion
Hashimoto's thyroiditis	Hypothalamic hypopituitarism
Thyroiditis with a low titer of autoantibodies	Immaturity of the hypothalamic–pituitary–thyroid axis
Transient autoimmune thyroiditis	Inappropriate secretion of TSH
Low reverse T_3 levels	

(serotonin, somatostatin, etc.) (central disorder), also related to proved hypothalamic dysfunction in these patients (13, 14).

The TRH test also reveals abnormal TSH secretion (delayed and prolonged response) in children with DS, believed to have 'normal' thyroid function, supporting the hypothesis of a central disorder of TSH secretion in some of them (15). It is possible that immaturity of the hypothalamic–pituitary–thyroid axis could cause the secretion of immunologically competent but biologically less active TSH in DS (15, 16). Table 5 shows the possible etiopathogenesis of high TSH levels in DS.

These thyroid abnormalities influence only in some cases the growth parameters of these patients (height, bone age, growth velocity), since other factors besides thyroid dysfunction could probably cause the shortness of stature that usually becomes more accentuated with increasing age in all patients with DS (9). Therefore, we suggest that all children with DS should undergo a careful periodical clinical and laboratory screening from the first months of life, in order that precocious thyroid functional disorders can be recognized, and the etiopathogenesis verified.

The finding of elevated basal TSH levels between 5 and 10 μU/ml and thyroid hormones within the normal range, even if they are considered high in a 'normal' subject, does not seem by itself to be a 'marker' of thyroid disease in children with DS. The possible transience and the 'fluctuations' in the follow-up of TSH values suggest a periodical control of thyroid function before embarking on substitutional therapy (L-thyroxine). This treatment will be necessary only for patients with persistently high TSH levels associated with clinical and auxological alterations.

ACKNOWLEDGMENTS

The authors wish to express particular gratitude to the nursing staff of the Ospedale Pediatrico 'Bambino Gesù' for their valuable help. This work was partly supported by grants from the Health Ministry, Rome, Italy.

REFERENCES

1. Sare Z, Ruvalcaba RHA, Kelley VC. Prevalence of thyroid disorder in Down's syndrome *Clin Genet* 1978; **14:** 154–158,
2. Cutler AT, Benezra-Obeiter R, Brink SJ. Thyroid function in young children with Down's syndrome, *Am J Dis Child* 1986; **140:** 479–483.
3. Colombo ML, Bona G, Quaglia P, Zaffaroni M, Maina D, Luotti D. Thyroid function in Down's syndrome children. *Minerva Pediatr* 1992; **44:** 11–16.
4. Fort P, Lifshitz F, Bellisario R, Davis J, Lanes R, Pugliese M, Richman R, Post EM, David R. Abnormalities of thyroid function in infants with Down's syndrome *J Pediatr* 1984; **104:** 545–549.
5. Ruch W, Schurmann K, Gordon P, Burgin-Wolf A, Girard J. Coexistent coeliac disease, Graves' disease and diabetes mellitus type 1 in patient with Down's syndrome. *Eur J Pediatr* 1985; **144:** 89–90.
6. Crinò A, Lucentini L, Ciolfi M, Matteoli MC, Cambiaso P, Giannone G, Zerella A, Borrelli P. Sindrome di Down e patologia autoimmune. In *La sindrome di Down*, edited by Ferri R, Spagnolo A, Il Pensiero Scientifico, Rome, 1989; pp. 203–206.
7. Hestnes A, Stovner LJ, Husoy O, Folling I, Fougner KJ, Sjaastad O. Hormonal and biochemical disturbances in Down's syndrome *J Ment Defic Res* 1991; **35:** 179–193.
8. Pueschel SM, Pezzullo JC. Thyroid dysfunction in Down's syndrome. *Am J Dis Child* 1985; **139:** 636–639.
9. Sharav T, Collins RM, Baab PJ. Growth studies in infants and children with Down's syndrome and elevated levels of thyrotropin. *Am J Dis Child* 1988; **142:** 1302–1306.
10. Crinò A, Giannotti A, Digilio MC, Albertini G, De Marchi C, Borrelli P. Evaluation of elevated levels of TSH in children and adolescents with Down's syndrome. *J Endocrinol Invest* 1990; **13 (suppl 3):** 76.
11. Cronk C, Croker AC, Pueschel SM, Shea AM, Zachai E, Pichens G, Reed RB. Growth charts for children with DS: 1 month to 18 years of age. *Pediatrics* 1988; **81:** 102–110.
12. Greulich WW, Pyle SI. *Radiographic Atlas of Skeletal Development of the Hand and Wrist*, 2nd edn, Stanford University Press, Stanford, CA, 1959.
13. Bhaumik S, Collacott RA, Garrick P, Mitchell C. Effect of thyroid stimulating hormone on adaptive behaviour in Down's syndrome. *J Ment Defic Res* 1991; **35:** 512–520.
14. Wisniewski KE, Bobinski M. Hypothalamic abnormalities in Down's syndrome. *Prog Clin Biol Res* 1991; **373:** 153–167.
15. Pinchera A, Martino E, Faglia G. Central hypothyroidism. In Werner's *The Thyroid: a Fundamental and Clinical Text*, edited by Ingbar SH, Brauerkan LE, Lippincott, New York, 1986; pp. 1235–1254.
16. Sharav T, Landau H, Zadik Z, Einarson TR. Age-related patterns of thyroid-stimulating hormone response to thyrotropin-releasing hormone stimulation in Down's syndrome. *Am J Dis Child* 1991; **145:** 172–177.

Part II

ANIMAL MODELS OF DOWN'S SYNDROME

10 The Trisomy 16 Mouse as an Animal Model Relevant to Studies of Growth Retardation in Down's Syndrome

MARY LOU OSTER-GRANITE
Division of Biomedical Sciences, University of California, Riverside, California, USA

INTRODUCTION

Short stature is considered a cardinal feature of Down's syndrome (DS), particularly in the adult period. Linear growth in children with DS is retarded, relative to normal children, beginning in prenatal life and extending through adolescence (1). Growth retardation of variable degree has been measured in body and organ weight, although placental weight appears to be unaffected during the second trimester (2). Linear growth of long bones appears to be affected by the second trimester of gestation, and such growth retardation has been suggested as one criterion for prenatal screening for DS, but its use as a diagnostic practice remains controversial (3). During postnatal life, growth retardation becomes exaggerated during periods of most rapid growth: the growth spurts of infancy and adolescence (4).

In addition to retarded linear growth, microcephaly is a common feature of DS (5). The concordance of clinical features suggests the possibility of an etiological relationship between brain dysfunction, microcephaly and growth retardation, since these children exhibit a reduced serum growth hormone (GH) response to L-dopa and clonidine stimulation tests, a disparity in responses, or a low integrated 24 h GH concentration (1). Thus the growth retardation observed in DS children may be due to a dysfunction of the hypothalamic–pituitary axis, secondary to diffuse neuropathological abnormalities (1), including neocortical cell reduction in layers II and IV of the visual cortex, abnormal development of the olfactory portions of the prefrontal cortex and auditory portions of the superior temporal gyrus, and reductions in cell populations in the arcuate and ventromedial nucleus (VMN) of the hypothalamus (Wisniewski *et al.*, this volume). It is the reduction of hypothalamic cell populations that may figure significantly in the proposed hypothalamo-hypophyseal target tissue dysfunction in DS individuals.

Growth Hormone Treatment in Down's Syndrome. Edited by S. Castells and K.E. Wisniewski

NORMAL HORMONAL INTERACTIONS DURING GROWTH

During the process of growth, secretion of GH is not chronic, but rather is episodic or pulsatile. This episodic secretion results from a rhythmic alternation of release of two hormones from cell populations present in the hypothalamus: those producing growth hormone-releasing hormone (GHRH), and those producing somatotropin release-inhibiting hormone (SRIF) or somatostatin (as reviewed recently by Devesa *et al.* (6)). These hormones are released into the portal circulation, where they interact with cells within the anterior pituitary—somatotrophs, lactotrophs and thyrotrophs—which secrete, respectively, GH, prolactin (PRL) and thyrotropin (TSH). The rhythm is controlled in large part by stimulating SRIF and by inhibiting GHRH. Alternations in the secretion of GHRH and SRIF set up a finely oscillating ultradian rhythm which helps to cause the fluctuations in the episodic release of GH—fluctuations whose magnitude is particularly important during the growth spurt of puberty (6).

Neurotransmitters modulate GH release, and perhaps the most important pathways identified to date are adrenergic pathways, although other neurotransmitters, such as acetylcholine, facilitate GH secretion by inhibiting release of SRIF expression in specific portions of the nervous system (6). Metabolic intermediates can also affect GH secretion: arginine and hypoglycemia inhibit SRIF release, whereas hyperglycemia and free fatty acids (FFAs) stimulate its release. In addition, circulating FFAs exert powerful inhibitory effects on the somatotrophs themselves (6). Other hormones, such as thyroid hormones, corticosteroids and sex steroids, act to modulate GH release (6). Sex steroids may exert their effects on the maturation of the growth homeostatis axis (and therefore modify the amplitude and frequency of GH release at puberty and in adulthood) through imprinting during fetal life (6). With the sex steroids, it is principally by indirect means that estradiol affects GH secretion (7), acting mainly to alter α_2-adrenergic transmission to SRIF cells (6). In general, these hormonal effects operate through modification of the functional state of ATP-sensitive K^+ channels; hormonal activation of GH release is inhibited by drugs which open ATP-sensitive K^+ channels and inhibition is suppressed by drugs which block ATP-sensitive K^+ channels (8).

In the periphery, GH stimulates the formation of insulin-like growth factors (IGF-1) or somatomedins which influence the growth of tissues directly during the postnatal period. In fetuses and neonates, IGF-2 is the predominant form of IGF. That the IGFs can influence fetal brain growth is evident from the expression of a binding protein, IGFBP-2, which is expressed in rat embryos as early as day 13.5–15 of gestation (E13.5–15) in the developing nervous system in choroid plexus epithelium, floor plate and infundibulum. IGF-2 itself is expressed in choroid plexus mesenchyme and Rathke's pouch progenitors, but not floor plate, so that interactions between the two may be extracellular or IGFBP-2 may have other functions (9). IGF-2, however, is not expressed in the neurectoderm or surface ectoderm during E8.5 and may act in an autocrine manner, since colocalization of IGF-2 transcripts, translated preproIGF-2 and the cognate IGF-2/mannose 6-phosphate receptor has been observed (10).

During the postnatal period, GH may have direct feedback effects on SRIF neurons, since GH receptors colocalize on SRIF neurons in the periventricular nucleus and paraventricular nucleus in the hypothalamus, but not on SRIF neurons in the cortex (11). These direct and indirect interactions of the varied hormones in the central nervous system, pituitary and periphery constitute a complex feedback mechanism which is responsible for growth homeostasis in the normal individual.

ROLE OF THYROID HORMONE DYSFUNCTION IN DOWN'S SYNDROME

Individuals with DS could have the process of growth homeostasis affected at various points along the pathway by alterations of hormone levels or of the location and distribution of their receptors and by imprinting effects during development, the latter effects exerted principally through effects of sex hormones. Historically, treatment with thyroid hormones was the first therapeutic intervention performed on mentally retarded children with growth deficiency (12). Thyroid hormones have been used to stimulate growth in individuals with growth retardation, since the lack of thyroid hormones strongly affects the hypothalamic synthesis of GHRH (13). Not only does this impair GH synthesis directly, but hypothyroidism exerts additional effects which depend on the lack of both positive modulation that thyroid hormones induce on the number of GHRH receptor sites in somatotrophs and the facilitator role of thyroid hormones on GHRH binding to its receptors (6). GH secretory mechanisms are also impaired in hypothyroidism (6).

It was Sutherland (12) who was able to use thyroxine (T_4) to ameliorate the effects of cretinism; his efforts with DS children proved less successful. Later, Benda found that administration of T_4 to DS individuals was not particularly helpful to DS individuals (Sharav, this volume). Koch *et al.* (14) had found that DS children proved not to be responsive to sodium lyothyronine (Cytomel); there was no effect on development, growth or height, or on general clinical condition. Finally, Sharav (this volume) found the effect of T_4 appeared to be compromised by increased TSH—increases which were variable among DS children as well as within in the same child at different ages.

Infants with DS who have growth retardation frequently have high circulating levels of TSH (15), which may in later life be associated with autoimmune thyroid disease (16). The genesis of these increased levels is not clear, but Sharav has cautioned that DS individuals may also have an immature hypothalamic response to GH (Sharav, this volume). Such an immature response may mean that needs are not met well by treatment with GH, but can be met better by treatment with IGF-1 (Sharav, this volume). IGF-1 maturation, however, may also be impaired, since Sara *et al.* (17) have found that incomplete switching from the fetal form of IGF-1 to GH-related IGF occurs in DS individuals. In older DS individuals, zinc treatment has proved helpful in reducing the incidence of subclinical hypothyroidism (18). Finally, individuals with DS who have manifestations of Alzheimer's disease are more likely to have autoimmune thyroiditis with the subclinical form of hypothyroidism, which may contribute to the cognitive defects in aging DS individuals (19). Thus at each major step of the path from hypothalamus to periphery there appears to be altered maturation and function of growth homeostasis in DS individuals.

NEUROCHEMICAL ABNORMALITIES IN DOWN'S SYNDROME INDIVIDUALS

With the realization that cell numbers in specific portions of the central nervous system appear to be affected in DS individuals (Wisniewski *et al.*, this volume), investigators have explored the status of various neurotransmitter systems. While norepinephrine (NE) levels may be elevated, cholinergic levels, serotoninergic levels and SRIF levels are reduced in DS individuals (Fabris *et al.*, this volume). Fabris *et al.* have suggested that neurotransmitter deficits, particularly those involving SRIF and serotonin, may be related to a zinc deficiency, suggesting dietary compromise, perhaps mediated by malabsorption-related malnutrition.

MALABSORPTION SYNDROME IN DOWN'S SYNDROME

DS patients do suffer from malabsorption, which may put them in a state of relative malnutrition (20). Hypercarotenemia occurs in children with DS and may reflect in part the increased amounts of unstable oxygen radical metabolites they form as well as decreased conversion of β-carotene to vitamin A (21). That increased oxygen radicals arise may not be surprising given the increases in superoxide dismutase and glutathione peroxidase, and decreases in catalase that have been observed (Fabris *et al.*, this volume). While Pueschel *et al.* (22) found that absorption of vitamin A was not decreased significantly in DS patients, vitamin therapy has not proved to be of great benefit to young children with DS (23). Malabsorption associated with gliadin antibodies to gluten has also been observed in DS children, suggesting a broad spectrum of malabsorption-related deficiencies (24). Thus, not only are there structural and functional alterations that are observed in the growth homeostasis axis, but confounding (or perhaps contributing) to these problems are the effects of compromised nutritional status.

THE TRISOMY 16 MOUSE AS A MODEL SYSTEM FOR STUDIES RELEVANT TO DOWN'S SYNDROME

In addition to significant growth retardation, DS represents the most common genetic cause of mental retardation in the human population (25). With LeJeune's discovery that DS was due to triplication of the smallest autosome—chromosome 21 (HSA 21)—the search for the specific genes involved, when triplicated, in generation of the characteristic features of DS began (26). Reasonably soon, however, it became clear that some individuals with DS inherited only a part of HSA 21 in triplicate, usually the distal third to half of the long arm of the chromosome (27). Since 1979, a large number of genes, DNA polymorphisms and other markers have been localized to HSA 21, and specific regions of the distal part of the chromosome appear to be associated with specific phenotypic characteristics when present in triplicate (28). Of the estimated 750–1500 genes present on HSA 21, about 30–40 have been identified and such identification is a major target of the human genome initiative (28). Interestingly, these genes appear to be conserved not only as a linkage group, but also in the same order, in other mammals, such as dogs, cats, cows and rodents (29).

In order to understand the pathogenesis of alterations that occur in the growth homeostasis axis in DS children, and the roles that specific genes and their overexpression play in that pathogenesis, a systematic analysis of the developmental interactions which occur among the various cell populations in brain, pituitary and periphery needs to be established. Unfortunately, critical events in early development cannot be examined in DS individuals, because they occur during the first trimester of gestation, when DS fetuses are neither grossly recognizable nor accessible. Therefore, it has become increasingly important to investigate the pathogenesis of alterations in brain and peripheral tissue development in animal model systems that express phenotypic characteristics of DS individuals (30–32).

One such animal model system, the trisomic mouse, was developed as a research tool by Alfred Gropp of Lubeck, Germany, during the 1970s. Using matings of mice with Robertsonian translocations creating fusions of specific chromosomes, Gropp was able to generate animals which contained a given chromosome in two copies, but both copies were attached to other chromosomes (as reviewed in Gearhart *et al.* (33)). When these animals were mated to genetically normal mice lacking Robertsonian chromosomes, trisomic and monosomic offspring were generated. Using this breeding strategy, Gropp was able to generate trisomy for each of the 19 autosomes in the mouse. His studies of the phenotypes displayed by the various trisomies led him to predict in 1974 that, based on phenotypic characteristics alone, trisomy 16 (Ts16) mice, with their virtually unique congenital heart defects, abnormal liver and hematopoietic development, and abnormal, yet characteristic, craniofacial features were the mice that most resembled the phenotypic features of DS (as reviewed in Gearhart *et al.* (33)).

It is to mouse chromosome 16 (MMU 16) that many of the genes in the 'DS region' of HSA 21 have been mapped (29). However, MMU 16 has proved to be about twice as large as HSA 21 and genes present on other human chromosomes (HSA 3, 17, 22, etc.) are also present on MMU 16. In addition, some genes present on the distal tip of the long arm of HSA 21 (q22.3→qter) are not present on MMU 16 (29), and these include some genes that have been implicated in features of the DS phenotype, such as a gene expressed in the nervous system in glial cells, *S-100β* (34). Two particularly important regions of MMU 16 whose triplication may produce alterations of growth homeostasis are located respectively on HSA 3 and probably on HSA 21 and encode the loci *preprosomatostatin* (*Smst*) and *dwarf* (*dw*).

THE PHENOTYPE OF THE TRISOMY 16 MOUSE

More systematic developmental studies of Ts16 mice since 1981 have demonstrated that these mice do possess many of the developmental deficits observed in DS children, including significant growth retardation (30–32). Within the central nervous system, for example, neurons of the cholinergic system—a vulnerable neurotransmitter system in DS children and adults—begin neurogenesis at the same time in both normal and trisomic mice, but cease proliferation prematurely so that the population of cholinergic neurons generated is reduced by about 40% by day 16 of gestation (E16) (35). Hypothalamic neuron populations are also reduced in cell number, although the patterns of neurogenesis of these cell populations has not been established (Wisniewski *et al.*, this volume). Finally, neurotransmitter alterations in noradrenergic, serotoninergic and cholinergic markers are observed in Ts16 mice, as early as neurogenesis in these populations has been completed (35, 36). These neurotransmitter abnormalities differ significantly from those of trisomy 19 (Ts19) mice (37), and resemble closely those observed in DS individuals.

Craniofacial development is similarly affected in Ts16 mice, producing significant developmental alteration and flattening of the facial skeleton, reduced angle of the skull base, and significant development of both the cochlea and semicircular canals in the audiovestibular system (32). With alterations in the development of the base of the skull, particularly in the sphenoid, basisphenoid and ethmoid bones, pituitary development is compromised (Oster-Granite, Ton and Pandya, unpublished data).

Growth retardation in the body of Ts16 mice is evident by mid-gestation, despite significant edema and hydrops (33), but neither the cellular basis of this retardation nor the time during gestation when it is first manifest has been determined. Generally, development of the anlagen of many of the endocrine glands, such as the thyroid gland, thymus and parathyroid glands in the head and neck and pancreas in the digestive system, is delayed or deficient by mid-gestation (Oster-Granite, unpublished observations). Thus compromised development of the central nervous system, the pituitary and the peripheral endocrine structures is evident by mid-gestation in the Ts16 mouse.

Despite these significant abnormalities, Ts16 mice are capable of responding to exogenous treatment with GH during the intrauterine period. Wisniewski *et al.* (this volume) have observed that administration of GH to litters containing Ts16 mice produces increased growth in both normal littermates and Ts16 fetuses—an effect she believes to be due in part to stimulation of IGF-1 production (Wisniewski *et al.*, this volume). Further, these authors find that the response to GH by Ts16 mice appears to be blunted—a potential indication of relative peripheral insensitivity to exogenous GH. Studies of IGF-1 levels or the effects of IGF treatment have not been described in Ts16 mice.

Wisniewski *et al.* (this volume) have obtained better success for growth in DS children with administration of GHRH, since the primary defect does not appear to be the ability to make functional GH, but the ability to generate substantial quantities of GH. One way to increase the effect of GHRH and thus the quantity of GH released by the pituitary would be to block the release of SRIF as well. In fact, augmented response is obtained in DS children when the release of SRIF is blocked through the administration of atenolol, which induces a β-adrenergic blockade, although this may be due in part to effects of SRIF on TSH secretion in the paraventricular nucleus (38). Finally, the inclusion in the therapeutic 'cocktail' of other fragments of the GH family of molecules, such as the GH polypeptides (GHPP), which are enkephalin-like peptides, produced a synergistic effect to improve the growth response (Wisniewski *et al.*, this volume). Neither atenolol nor GHPP has been administered to Ts16 mice during gestation.

Unfortunately, studies of growth regulation in the postnatal period cannot be accomplished in Ts16 mice because Ts16 mice do not survive birth. This has necessitated the development of other strategies, such as the generation of transgenic mice and mice with portions of MMU 16 in triplicate (segmental aneuploidy) to study further these important relationships in growth homeostasis. To understand the importance of the observation that Ts16 mice respond to GH by growth during the intrauterine period, it is helpful to discuss current opinions about the ontogenesis of GH secretion by the anterior pituitary in rodents, a field which has expanded significantly over the past five years, and the revelations that have appeared from studies of mutant mice with congenital dwarfism.

GROWTH RETARDATION ASSOCIATED WITH EXPRESSION OF THE *dwarf* LOCUS IN MICE

The availability of genetic models of growth retardation has led to the study of a variety of mice which exhibit single-locus mutations producing dwarfism of varying severity (39). One of the most interesting groups of mutations are three dwarf mutations: Snell, Jackson and Ames. Both the Snell and Jackson dwarf mice result from mutations of a gene located on MMU 16, *dwarf* (*dw*). Study of such mutants may be relevant to the situation in DS, since it is not clear at present where the proximal boundary of synteny or homology between MMU 16 and HSA 21 lies along MMU 16. The gene product of the *dw* locus is, thus, overexpressed in Ts16 mice and may be overexpressed in DS individuals as well. By its overexpression, *dw* may alter pituitary and/or hypothalamic development to contribute to the growth retardation observed in Ts16 mice.

Homozygous *dw* mice have some of the characteristic features observed in DS children. First, they cannot be distinguished easily by length or head circumference from their normal littermates at birth. During infancy (the first seven postnatal days), a progressive deceleration of growth occurs which becomes greater as these mice enter adolescence (days 14–21); significant reduction in body length (60%) and in brain size (35%) is observable by day 21 postnatal (39). In addition, these mice exhibit hypothyroidism which complicates their postnatal growth and brain development, and hypogonadism which renders them sterile (39). Exogenous supplementation of homozygous *dw* mice with T_4, GH and PRL restores growth as well as fertility to female mice, but has been a rather unsuccessful therapy for bringing about fertility in male mice (Oster-Granite, unpublished data). Although pituitary development is disrupted in Snell *dw* mice (40), no systematic analysis of litters containing prenatal homozygous *dw* mice has been attempted because litters at best contain only 50% homozygous *dw* progeny, since they result from the mating of a homozygous hormonally supplemented female with a heterozygous male, and the homozygous and heterozygous *dw* progeny are unrecognizable during the intrauterine period. Critical to such prenatal analyses, then, is the identification of the product of the *dw* locus and identification of its alteration in *dw* mice.

Our previous studies of *dw* mice (40) were inspired by two observations: *dw* mice had (a) cholinergic deficits and increased levels of SRIF in their brains (41, 42), and (b) abnormal development of cortical neurons in area 6 (43). Thus *dw* mice shared three characteristic features we had previously observed in Ts16 mice: overexpression of *Smst* (44), cholinergic deficits of undefined origin (36), and abnormal neurogenesis. Since both *dw* and the gene encoding preprosomatostatin (*Smst*) were located on the proximal portion of MMU 16, they might be alleles. Our subsequent mapping studies (40) revealed that these genes were not allelic. Although there was overexpression of *Smst* message in the cortex of *dw* mice by *in situ* hybridization, *Smst* message levels were decreased in hypothalamus of the same adult animals (44, 45). Since *dw* and *Smst* were not alleles, further studies of *dw* mice had to await the determination of the gene product of the locus. This product has recently been identified and its identification has helped to explain the correlation of altered GH, PRL and TSH function in *dw* mice, and perhaps in DS individuals as well.

THE ROLE OF *Pit-1* IN DEFECTS OBSERVED IN GROWTH IN THE TRISOMY 16 MOUSE

A critical question in anterior pituitary development has been what triggers the appearance of each of the cell types present in the adult: somatotrophs, lactotrophs, thyrotrophs, corticotrophs and gonadotrophs. These varied cell populations arise from an apparently homogeneous primordium, which appears as a diverticulum of the ectoderm of the oral cavity (Rathke's pouch) at E8.5 in the mouse (46). By E12 the diverticulum has detached from the oral epithelium and proliferation of the cells produces obliteration of its lumen. By E15–16 cells within this mass have associated themselves with a similar diverticulum from the diencephalon (which forms the posterior pituitary) and can now be identified by *in situ* hybridization or immunocytochemistry in some cases, as cells producing particular hormones.

Since GH and PRL are closely related genes (46), it is important to determine whether GH and PRL are produced simultaneously by young anterior pituitary cells or whether the cells have separate lineages. Expression of GH in cells begins at E16 in mouse and around E18 in rat (47). PRL, however, is not detectable by immunocytochemistry until postnatal day (PND) 10–12 in the mouse, although mice transgenic for the *Prl* gene do express PRL message during the late fetal period (48, 49). Most PRL cells appear to arise from a cohort of cells which initially produce GH and PRL messages and which then become limited in expression to one or the other. Thus GH and PRL cells appear to have a common precursor lineage. That a few cells can express PRL in the absence of expressing GH is evident from studies of transgenic mice in which the GH gene has been deleted by introduction of a diphtheria toxin vector construct (50). The final number of cells expressing GH and PRL within the anterior pituitary appears to depend on imprinting by sex steroids during a 'critical' period of intrauterine life. Those animals experiencing high levels of testosterone are destined to have more cells expressing GH than PRL, and those not exposed to high levels of testosterone have about equal numbers of cells expressing GH and PRL (6).

Initiation of expression of GH, PRL and TSH occurs about 48 h after the appearance of expression of a pituitary specific transcription factor (GHF–1; *Pit-1*) (49). *Pit-1* contains two highly conserved amino acid sequence motifs: a 75 amino acid sequence motif, the POU-specific domain, which is shared with *Oct-1*, *Oct-2* and *Unc-86* (hence the name POU), and a 60 amino acid homeodomain sequence motif at the carboxyl end, which is involved in DNA binding, and thus activation of GH transcription. While *Pit-1* expression appears early and transiently in the rat neural tube and embryonic brain, in mouse it seems restricted to the pituitary, although other members of the POU domain family are expressed in mouse brain at the same gestational stages (51). Some investigators believe that the expression of *Pit-1* is likely to represent the very last step in the specialization pathway leading to formation of fully determined somatotrophs (46).

In human beings, mutations of *Pit-1* in the DNA binding region are associated with combined pituitary deficiency (52), but mutations which disrupt the structure of the POU domain are not (53). Snell and Jackson *dw* mice probably result from mutations in *Pit-1*, since there is a lack of significant recombination of the *Pit-1* locus with the *dw* locus (54). In the case of the Jackson *dw*, the mutation in *Pit-1* rearranges the gene, and in the case of the Snell *dw*, the mutation produces missense in the homeodomain region, which eliminates the binding of the defective protein to the GH and PRL genes (55).

Although *Pit-1* expression correlates functionally with that of the GH and PRL genes, its expression may also be necessary for the survival of thyrotrophs, since the introduction of antisense into pituitary line cultures is able to decrease proliferation of thyrotrophs (56). Finally, the gene product of the Ames *dwarf* mutation, located on mouse chromosome 11, may function epistatically with *Pit-1* (46). Functionally, then, when *Pit-1* is absent, the inactive promoter of GH is refractory to glucocorticoid, thyroid hormone receptors and cAMP response factors (46). In the case of the Snell and Jackson *dw* mice, the failure of *Pit-1* expression may result in failure of formation of GH and PRL cells and may decrease the survival of TSH cells (39). In the case of the Ts16 mouse, overexpression of *Pit-1*, since there are three copies of the gene due to triplication of chromosome 16, may produce similar effects.

Although *Pit-1* is overexpressed in the Ts16 mouse (Oster-Granite, Ton and Pandya, unpublished observations), the impact of its overexpression at very early stages of pituitary development cannot be assessed during the important postnatal growth period in Ts16 mice, since none of them survives birth. It is, however, intriguing that a gene whose expression is not detected in regions of the mouse brain during early development, such as the cortex or basal forebrain, has associated with its mutation alterations in the expression of two neurotransmitter systems whose function can impact directly upon growth homeostasis: the SRIF and cholinergic systems. To ascertain the relationship between absence of gene function and increased gene dosage, and thus to further study pituitary development relevant to the effects of *Pit-1*, transgenic mice with multiple expressed copies of the *Pit-1* gene must be generated. Mice transgenic for *Pit-1* have not yet been described. Thus growth effects on pituitary development in Ts16 mice may be the result of overexpression of a transcription factor which alters GH, PRL and TSH cell development in the pituitary directly, and/or they may result from alterations in SRIF and cholinergic function that occur when this locus is altered in expression.

THE *Smst* GENE AND ITS EXPRESSION IN TRISOMY 16 MICE

Another gene whose overexpression could impact upon growth in the intrauterine period in the mouse is *preprosomatostatin* (*Smst*). As discussed above, SRIF modulates GH and PRL secretion during normal patterns of growth. Additionally, SRIF appears transiently during neurogenesis in cortex, brain stem and cerebellum, and may modulate mitotic activity of cells (57). One might anticipate, certainly during the postnatal period, that overexpression of *Smst* would lead to increased levels of SRIF which might in turn decrease GH and ultimately IGF-1 production, and thus lead to dwarfism.

In Ts16 mice, overexpression of *Smst* can be detected as early as E11 (44), at a time when overexpression of *Smst* in the developing nervous system might lead to premature cessation of cell division. Given the pivotal role that SRIF plays in growth homeostasis, the generation of mice with excess copies solely of the *Smst* gene should provide animals that survive into the postnatal period and potentially that would manifest both growth abnormalities and neurochemical abnormalities since there is a dynamic interaction between cholinergic function and SRIF function during neocortical neurogenesis (58).

Smst TRANSGENIC MICE AND THEIR CHARACTERISTICS

Our group generated *Smst* transgenic (*Smst*Tg) mice by introducing a 15 kb genomic fragment containing the endogenous mouse *Smst* gene—both exons and intervening intron and 9 kb of 5′ flanking sequence (59, 60), a genomic construct of sufficient size to contain all needed promoter and enhancer elements for us to obtain stage- and tissue-specific expression (61). From six founder ($B6AF_1 \times C57BL/6J$) mice, three lines of *Smst*Tg mice were established: *Smst*Tg13, *Smst*Tg16, and *Smst*Tg42. In each of these lines, multiple copies of the transgene were integrated as concatemers, producing different and distinctive patterns on Southern blots (62). In progeny from *Smst*Tgl6 and *Smst*Tg42, gene copy was maintained in subsequent generations of intercrosses with C57BL/6J mice, and in *Smst*Tg13 two sites of integration occurred which were segregated by mating. One line of segregants proved infertile and was subsequently lost. Lines *Smst*Tg13 and *Smst*Tg42 can be maintained at homozygosity of both males and females in the absence of the effects of genomic imprinting, but line *Smst*Tg16 has proven more problematic; only occasionally can homozygous males, but never homozygous females, be obtained from hemizygous matings.

Analysis of *Smst* message levels revealed increases that were not homogeneous throughout the brain. Rather, levels of expression varied in different brain regions among adult transgenic mice of all three transgenic lines. Additionally, there appeared to be a sexual dimorphism in message levels (62). Overexpression of *Smst* message also occurred in fetal transgenic mice (60, 62), with a distribution similar to that observed in Ts16 mice (Oster-Granite and Bendotti, unpublished observations). Overexpression was accompanied by

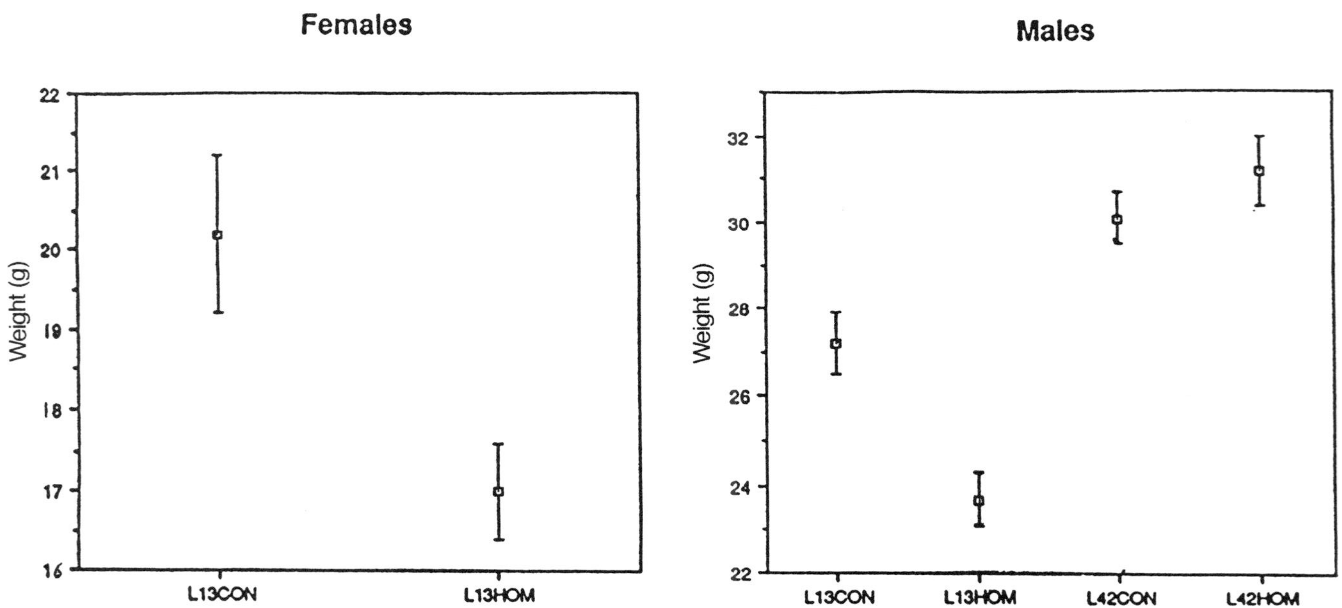

Figure 1. Comparison of body weights in adult somatostatin transgenic mice. Values are mean ± SEM from individuals body weights in grams. Transgenic values differ significantly from same line non-transgenic control animals by ANOVA ($p < 0.05$). L13CON = *Smst*Tg13 non-transgenic mice; L13HOM = *Smst*Tg13 homozygous transgenic mice; L42CON = *Smst*Tg42 non-transgenic mice; L42HOM = *Smst*Tg42 homozygous transgenic mice. Note that there is significant sexual dimorphism in size of *Smst*Tg13 mice, the average weights in female non-transgenic and transgenic respectively being 20 and 17 g and in male non-transgenic and transgenic mice 27 and 23.5 g

increased levels of SRIF-like immunoreactivity as revealed by radioimmunoassay (62) and increased numbers of SRIF-like immunoreactive cells (66).

When *Smst*Tg13 and *Smst*Tg42 mice were studied for a variety of spontaneous behaviours, hyperactivity appeared in the *Smst*Tg13 animals between four and six months of age, and *Smst*Tg42 animals also tended towards hyperactivity by six months of age. Unfortunately, the results were highly variable, again suggesting that sexual dimorphism may play a role in the behavioural effects observed (62). In initial neurochemical studies of *Smst*Tg13 mice, alterations were found in patterns of cholinergic markers that differed from those described for *dw* mice (63), but insufficient numbers of animals have been studied to ascertain whether the variations observed are due to sexual dimorphism in particular brain regions of the transgenic mice.

Patterns of growth were studied in these three lines of transgenic mice. Body weights of adult homozygous *Smst*Tg13 mice were significantly smaller (10%) than their non-transgenic controls, and they resemble *dw* mice in size and weight at four months of age (Fig. 1). In contrast, homozygous *Smst*Tg42 animals were not smaller than their non-transgenic controls (Fig. 1). When similar measurements of length (Fig. 2) and weight (Fig. 3) were made on litters of transgenic mice at E15, *Smst*Tg42 mice, whether hemizygous or

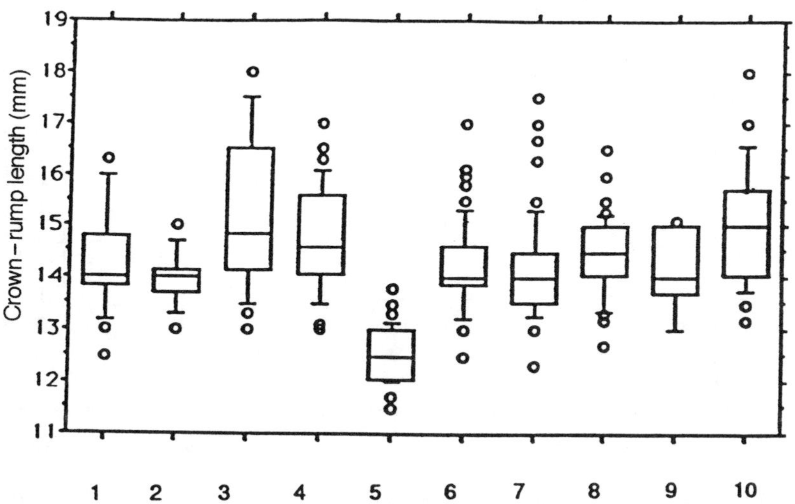

Figure 2. Comparison of the crown–rump lengths (in millimeters) of litters of mice at E15. 1 = *Smst*Tg13 non-transgenic mice; 2 = *Smst*Tg42 non-transgenic mice; 3 = *Smst*Tg16 non-transgenic mice; 4 = C57BL/6J mice; 5 = A/J mice; 6 = *Smst*Tg13 hemizygous transgenic mice; 7 = *Smst*Tg13 homozygous transgenic mice; 8 = *Smst*Tg42 hemizygous transgenic mice; 9 = *Smst*Tg42 homozygous transgenic mice; 10 = *Smst*Tg16 hemizygous transgenic mice. Note that the A/J mice appear smaller at E15 and that there is considerable variability in the size of various groups of mice

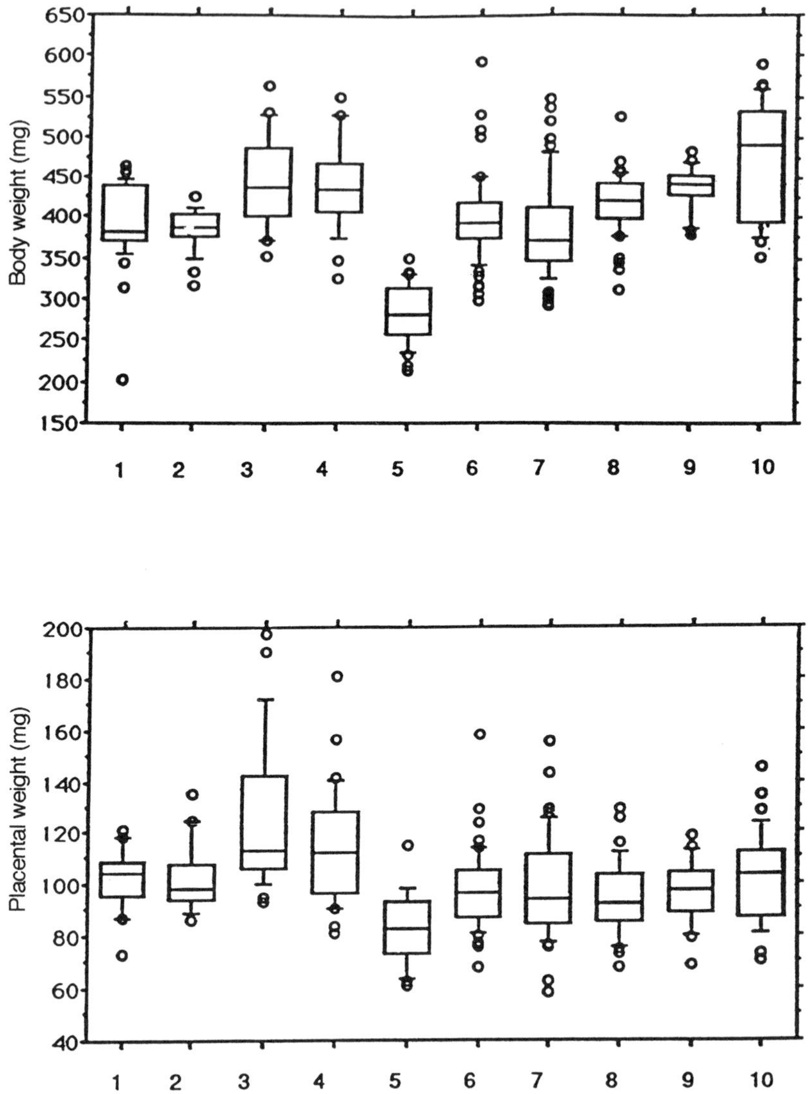

Figure 3. Comparison of body and placental weights (in milligrams) of litters of mice at E15. 1 = *Smst*Tg13 non-transgenic mice; 2 = *Smst*Tg42 non-transgenic mice; 3 = *Smst*Tg16 non-transgenic mice; 4 = C57BL/6J mice; 5 = A/J mice; 6 = *Smst*Tg13 hemizygous transgenic mice; 7 = *Smst*Tg13 homozygous transgenic mice; 8 = *Smst*Tg42 hemizygous transgenic mice; 9 = *Smst*Tg42 homozygous transgenic mice; 10 = *Smst*Tg16 hemizygous transgenic mice. Note that the A/J mice appear smaller at E15 and that there is considerable variability in the size of various groups of mice

homozygous and *Smst*Tgl6 hemizygous mice tended to be heavier than the *Smst*Tgl3 mice. In these experiments, sex of the conceptuses studied was not determined and significant variation was observed, so dimorphism may exist even at this early gestational age. Additionally, parental A strain mice were consistently and significantly ($p < 0.05$) smaller in length and weight at E15, even though they were much larger mice as adults (data not shown). The impact of the contribution of genes from the A strain to growth dynamics has not been determined in these mice, and efforts to breed them onto the C57BL/6J background continue. Thus one line of *Smst* transgenic mice is smaller and continues to remain smaller after E15 (*Smst*Tg13), and another line of transgenic mice (*Smst*Tg42) tend, if anything, to be larger. Current studies focus on identifying the insertion site of the transgene in these two lines, and determination of sex by polymerase chain reaction using X- and Y-specific probes in the fetuses.

Other investigators have produced similar *Smst* transgenic mice. Rubenstein *et al.* (64) used a construct containing both exons and intervening intron, 12 kb 5′ flanking region and 2 kb 3′ flanking region, a construct similar to our own construct, but with the addition of a 30mer transgene tag, which permitted them to identify unequivocally expression due to the transgene. They observed that areas with high levels of endogenous *Smst* mRNA were devoid of the transgene coded expression. Unexpectedly, they found that transgene expression also occurred in paraventricular, supraoptic and retrochiasmatic nuclei. Expression was colocalized with vasopressin, but not oxytocin, in magnocellular neurons. They interpreted this transgene expression in paraventricular and supraoptic nuclei to be ectopic, indicating that all the appropriate elements controlling *Smst* gene expression were not present in the transgene construct.

When *Smst* message distribution was examined in our adult *Smst*Tg13 hemizygous mice (Oster-Granite, Kinsman and Bendotti, unpublished data), the pattern of distribution did not differ significantly from that in nontransgenic mice (45), but there was a persistent hybridization signal in the habenular and supraoptic nuclei. The paraventricular, supraoptic, retrochiasmatic and habenular nuclei normally express *Smst* message transiently during the first 21 days of postnatal life (45). In our transgenics, this expression seems to persist into the adult period. Current studies focus on analysis of these lines of transgenic mice by *in situ* hybridization at various gestational and postnatal ages to ascertain whether increased expression is in fact persistent expression.

Persistence of *Smst* expression in the transgenic mice may be tied to the expression of SRIF receptors, and may indicate autocrine regulation of SRIF protein levels in these cell populations. Studies of SRIF receptor expression in mouse brain have led to the identification of at least two subtypes: SRIFR1 and SRIFR2 (65). SRIFR1 mRNA was found throughout the mouse brain, but particularly in supra- and infragranular layers of the cortex, the amygdala, hippocampus, bed nucleus of stria terminalis, substantia nigra, parabrachial nucleus and nucleus of the solitary tract. Within the hypothalamus, SRIFR1 receptors were abundant in medial preoptic nucleus and suprachiasmatic nucleus, as well as in the arcuate and posterior paraventricular nucleus. SRIFR2 receptors were detected only in the medial parvicellular subnucleus of the paraventricular nucleus, in the arcuate and posterior periventricular nucleus, and in the retrochiasmatic area. SRIFR2 receptors were found in the infragranular layers of cortex, the amygdala, the claustrum, endopiriform nucleus, the arcuate and paraventricular nuclei of the hypothalamus, and medial habenular nucleus. Several regions reported to contain dense SRIF-LI terminal fields and receptor binding sites were devoid of both SRIFR1 and SRIFR2 mRNA, so additional receptors types may exist (65). The colocalization of these receptors with regions overexpressing *Smst* may be similar to the IGF system discussed previously.

CONCLUSIONS

Ts16 mice provide an experimental system in which growth in the intrauterine period can be examined easily. In order for growth to be studied during the postnatal period, transgenic mice must be generated that overexpress particular genes triplicated in Ts16 mice. At least two genes located on MMU 16—*Smst* and *dw*—may be manipulated to generate mice which share a number of features of growth retardation with children with DS. In these mice alterations in SRIF and cholinergic neurotransmitter systems accompany alterations in pituitary development which compromise the generation and/or survival of cells producing GH, PRL and TSH. Studies of the genesis of alterations in growth, as well as the effects of specific therapeutic regimes, may provide further understanding of the basic mechanisms by which growth occurs in normal children, and may provide insight into the design of therapeutic regimes which can better effect growth in DS children.

REFERENCES

1. Castells S, Torrado C, Bastian W, Wisniewski KE. Growth hormone deficiency in Down's syndrome children. *J Intell Disability Res* 1992; **36:** 289–293.
2. Fitzsimmins J, Droste S, Shephard TH, Pascoe-Mason J, Fantel A. Growth failure in second trimester fetuses with trisomy 21. *Teratology* 1990; **42:** 337–345.
3. Nyberg DA, Resta RG, Hickok DE, Hollenbach KA, Luthy DA, Mahony BS. Femur length shortening in detection of Down's syndrome: Is prenatal screening feasible? *Am J Obstet Gynecol* 1990; **162:** 1247–1252.
4. Cronk C, Crocker AC, Pueschel SM, Shea AM, Zackai E, Pickens G, Reed RB. Growth charts for children with Down's syndrome: 1 month to 18 years of age. *Pediatrics* 1988; **81:** 102–110.

5. Wisniewski KE. Down's syndrome (DS) children postnatal retardation of head and brain growth (birth–five years). In *Disorders of the Developing Nervous System: Changing Views on their Origin, Diagnoses, and Treatment. Birth Defects Symposium XVIII*, edited by Swann W, Messer A, Liss, New York, 1988; pp. 183–194.
6. Devesa J, Lima L, Tresguerres JAF. Neuroendocrine control of growth hormone secretion in humans. *Trends Endocrinol Metab* 1992; **3:** 175–183.
7. Ho KY, Evans WS, Blizzard RM. Effects of sex and age on 24-hour profile of growth hormone secretion in man: importance of endogenous estradiol concentration. *J Clin Endocrinol Metab* 1987; **64:** 51–53.
8. deWeille JR, Fosset M, Epelbaum J, Lazdunski M. Effectors of ATP-sensitive K^+ channels inhibit the regulatory effects of somatostatin and GH-releasing factor on growth hormone secretion. *Biochem Biophys Res Commun* 1992; **187:** 1007–1014.
9. Wood TL, Brown AL, Rechler MM, Pintar JE. The expression pattern of an insulin-like growth factor (IGF)-binding protein gene is distinct from IGF-II in the midgestational rat embryo. *Mol Endocrinol* 1990; **4:** 1257–1263.
10. Lee JE, Pintar J, Efstratiadis A. Pattern of insulin-like growth factor II gene expression during early mouse embryogenesis. *Development* 1990; **130:** 151–159.
11. Burton KA, Kabigting EB, Clifton DK and Steiner RA. Growth hormone receptor messenger ribonucleic acid distribution in the adult male rat brain and its colocalization in hypothalamic somatostatin neurons. *Endocrinology* 1992; **131:** 958–963.
12. Sutherland (1900) quoted in Smith GF and Berg DM. *Down's Anomaly*. 2nd Edition. 1965. Churchill Ltd., London.
13. Katakami H, Downs TR, Frohmann LA. Inhibitory effect of hypothalamic preoptic area somatostatin on growth hormone releasing hormone in the rat. *Endocrinology* 1988; **123:** 1103–1109.
14. Koch R, Share J, Graliker B. The effect of cytomel on young children with Down's syndrome (mongolism): a double blind longitudinal study. *J Pediatr* 1965; **66:** 776–778.
15. Sharav T, Landau H, Zadik Z, Einarson TR. Age-related patterns of thyroid stimulating hormone response to thyrotropin-releasing hormone stimulation in DS children. *Am J Dis Child* 1991; **145:** 172–175.
16. Dianani S, Carpenter S. Down's syndrome and thyroid disorder. *J Ment Defic Res* 1991; **34:** 187–193.
17. Sara V, Gustavson KH, Anneren G, Hall K, Wetterberg L. Somatomedins in Down's syndrome. *Biol Psychiatry* 1983; **18:** 803–811.
18. Napolitano G, Palka G, Lio S, Bucci I, deRemigis P, Stuppia L, Monaco F. Is zinc deficiency the cause of subclinical hypothyroidism in Down Syndrome? *Ann Genet* 1990; **33:** 9–15.
19. Percy ME, Dalton AJ, Markovic VD, Crapper McLachlan DR, Gera E, Hummel JT, Rusk ACM, Somerville MJ, Andrews DF, Wlafish PG. Autoimmune thyroiditis associated with mild 'subclinical' hypothyroidism in adults with DS: a comparison of patients with and without manifestations of Alzheimer's disease. *Am J Med Genet* 1990; **36:** 148–154.
20. Abalan F, Jouan A, Weerts MT, Solles C, Brus J, Sauneron JF. A study of digestive absorption in four cases of Down's syndrome: Down's syndrome, malnutrition, malabsorption, and Alzheimer's disease. *Med Hypotheses* 1990; **31:** 35–38.
21. Storm W. Hypercarotenaemia in children with Down's syndrome. *J Ment Defic Res* 1990; **34:** 283–286.
22. Pueschel SM, Hillemeier C, Caldwell M, Senft K, Mevs C, Pezzulo JC. Vitamin A gastrointestinal absorption in persons with Down's syndrome. *J. Ment Defic Res* 1990; **34:** 269–275.

23. Preuss JB, Fewell RR, Bennett FC. Vitamin therapy and children with Down's syndrome: a review of research. *Except Child* 1989; **55:** 336–341.
24. Storm W. Prevalence and diagnostic significance of gliadin antibodies in children with Down's syndrome. *Eur J Pediatr* 1990; **149:** 833–834.
25. de la Cruz FF, Oster-Granite ML. Neural bases of mental retardation. In *Handbook of Human Biological Development*, Vol. 3, edited by Timiras PS, Meisami E, CRC Press, Boca Raton, FL, 1989; pp. 1–20.
26. Coyle JT, Oster-Granite ML, Gearhart JD. The neurobiologic consequences of Down's syndrome. *Brain Res Bull* 1986; **16:** 773–787.
27. Reeves RH, Morgan RA, Bendotti C, Oster-Granite ML, Coyle JT, Gearhart JD. The mouse as a model for studies of Alzheimer's disease. In *The Molecular Biology of Alzheimer's Disease. Banbury Conference 28*, edited by Davies P, Finch C, Cold Spring Harbor Laboratory, Cold Spring Harbor, 1988; pp. 123–128.
28. Epstein CJ, Korenberg JM, Anneren G, Antonarakis SE, Ayme S, Courchesne E, Epstein LB, Fowler A, Groner Y, Huret JL, Kemper TL, Lott IT, Lubin BH, Magenis E, Opitz JM, Patterson D, Preist JH, Pueschel SM, Rapoport SI, Sinet PM, Tanzi RE, de la Cruz F. Protocols to establish genotype–phenotype correlations in Down's syndrome. *Am J Hum Genet* 1991; **49:** 207–235.
29. Reeves RH, Miller RD. Mouse chromosome 16. *Mammalian Genome* 1992; **3:** S233–S240.
30. Oster-Granite ML, Reeves RH, Gearhart JD, Coyle JT. Disorders of the developing nervous system in the trisomy 16 mouse: relevance to Down's syndrome and Alzheimer's Disease. In *Disorders of the Developing Nervous System: Changing Views of Their Origins, Diagnoses, and Treatment.* Birth Defects Symposium XVIII, edited by Swann W, Messer A. Liss, New York, 1988; pp. 169–188.
31. Coyle JT, Oster-Granite ML, Reeves R, Hohmann C, Corsi P, Gearhart J. Down's syndrome and the trisomy 16 mouse: impact of gene imbalance on brain development and aging. In *Genes, Brain, and Behavior*, edited by McHugh PR, McKusick VA, Raven Press, New York, Research Publications of the Association of Research in Nervous and Mental Diseases 69, 1991; pp. 85–99.
32. Grausz H, Richtsmeier J, Oster-Granite ML. Morphogenesis of the brain and craniofacial complex in trisomy 16 mice. In *Morphogenesis of Down's Syndrome*, edited by Epstein CJ, Wiley–Liss, New York, 1991; pp. 169–188.
33. Gearhart JD, Davisson MT, Oster-Granite ML. Autosomal aneuploidy in mice: generation and developmental consequences. *Brain Res Bull* 1986; **16:** 789–801.
34. Reeves R, Hilt D, Moran T, Oster-Granite ML, Gearhart J. S100β transgenic mice: characterization and behavioral consequences of overexpression. *Int Hum Genome Meeting*, 1991; 20.
35. Sweeney JE, Höhmann CF, Oster-Granite ML, Coyle JT. Neurogenesis of the basal forebrain in normal and trisomy 16 mice: an animal model for developmental disorders in Down's syndrome. *Neuroscience* 1989; **31:** 413–425.
36. Ozand PT, Hawkins RL, Collins RM Jr, Reed WD, Baab PJ, Oster-Granite ML. Neurochemical changes in murine trisomy 16: delay in cholinergic and catecholaminergic systems. *J Neurochem* 1984; **43:** 401–408.
37. Saltarelli MD, Forloni GL, Oster-Granite ML, Gearhart JD, Coyle JT. Neurochemical characterization of embryonic brain development in trisomy 19 (Ts19) mice: implications of selective deficits observed for abnormal neuronal development in aneuploidy. *Dev Genet* 1987; **8:** 267–279.
38. Liau N, Vaudry H, Pelletier G. Neuroanatomical connections between corticotropin-release factor (CRF) and somatostatin (SRIF) nerve endings and thyrotropin-releasing hormone (TRH) neurons in the paraventricular nucleus of the rat hypothalamus. *Peptides* 1992; **13:** 677–680.

39. Lyon MF, Searle AG. *Genetic Variants and Strains of the Laboratory Mouse*, 2nd edn, Oxford University Press, New York, 1990; 876 pp.
40. O'Hara BF, Bendotti C, Reeves RH, Oster-Granite ML, Gearhart JD, Coyle JT. Genetic mapping and analysis of somatostatin expression in Snell dwarf mice. *Mol Brain Res* 1988; **4:** 283–292.
41. Fuhrmann G, DiScala-Guenot D, Ebel A. Somatostatin levels in the central nervous system of the Snell dwarf mouse: is somatostatin excess the primary molecular defect in the dw/dw dwarfism? *Brain Res* 1985; **328:** 161–164.
42. Fuhrmann G, Durkin T, Thiriet G, Kempf E, Ebel A. Cholinergic neurotransmission in the central nervous system of the Snell dwarf mouse *J Neurosci Res* 1985; **13:** 417–430.
43. Noguchi T, Sugisaki T, Takamatsuk, Tsukada Y. Neurochemical abnormalities for Snell dwarf mice. In: *Genetic Approaches to Developmental Neurobiology* edited by Tsukada Y, University of Tokyo Press, Tokyo 1982 pp. 153–170.
44. O'Hara BF, Fisher O, Oster-Granite ML, Gearhart JD, Reeves RH. Developmental expression of the amyloid precursor protein, growth associated protein 43, and somatostatin in normal and trisomy 16 mice. *Dev Brain Res* 1989; **49:** 300–304.
45. Bendotti C, Höhmann CF, Forloni GL, Reeves RH, Coyle JT, Oster-Granite ML. Developmental expression of somatostatin in mouse brain. II. *In situ* hybridization studies. *Dev Brain Res* 1990; **53:** 26–39.
46. Voss JW, Rosenfeld MG. Anterior pituitary development: short tales from dwarf mice. *Cell* 1992; **70:** 527–530.
47. Nagata S, Rosenfeld MG, Inoue K. Development of prolactin and growth hormone production in fetal rat pituitary: an immunochemical study. *Dev Growth Differentiation* 1992; **34:** 473–478.
48. Karin M, Castrillo JL, Theill LE. Growth hormone gene regulation: a paradigm for cell-type specific gene activation. *Trends Genet* 1990; **6:** 92–96.
49. Crenshaw EB III, Kalla K, Simmons DW, Swanson LW, Rosenfeld MG. Cell-specific expression of prolactin gene in transgenic mice is controlled by synergistic interactions between enhancer and promoter elements. *Genes Dev* 1989; **3:** 959–972.
50. Dolle P, Castrillo JL, Theill LE, Deerinck T, Ellisman M, Karin M. Expression of GHF-1 protein in mouse pituitaries correlates both temporally and spatially with the onset of growth hormone gene activity. *Cell* 1990; **60:** 809–820.
51. Ingraham HA, Albert VR, Chen R, Crenshaw EB III, Elsholtz HP, He X, Kapiloff MS, Mangalam HJ, Swanson LW, Treacy MN, Rosenfeld MG. A family of POU-domain and PIT-1 tissue-specific transcription factors in pituitary and neuroendocrine development. *Annu Rev Physiol* 1990; **52:** 773–791.
52. Radovick S, Nations M, Du Y, Berg LA, Weintraub BD, Wondisford FE. A mutation in the POU-homeodomain of Pit-1 is responsible for combined pituitary hormone deficiency. *Science* 1992; **257:** 1115–1118.
53. Pfäffle RW, DiMattia GE, Parks JS, Brown MR, Wit JM, Jansen M, van der Nat H, van den Brande JL, Rosenfeld MG, Ingraham HA. Mutation of POU-specific domain of Pit-1 and hypopituitarism without pituitary hypoplasia. *Science* 1991; **257:** 1118–1121.
54. Camper SA, Saunders TL, Katz RW, Reeves RH. The Pit-1 transcription factor gene is a candidate for the murine Snell dwarf mutation. *Genomics* 1991; **8:** 586–590.
55. Li S, Crenshaw EB, Rawson EJ, Simmons DM, Swanson LW, Rosenfeld MG. Dwarf locus mutants lacking three pituitary cell types result from mutations in the POU-domian gene Pit-1. *Nature* 1990; **347:** 528–530.

56. Castrillo JL, Theill LE, Karin M. Function of the homeodomain protein GHF-1 in pituitary cell diflerentiation. *Science* 1991; **253:** 197–199.
57. Daikoku S, Hisano S, Kawano H, Okamura Y, Tsuruo Y. Ontogenetic studies of topographic heterogeneity of somatostatin-containing neurous in the rat hypothalamus *Cell Tics Res* 1983; **233**: 347–354.
58. Höhmann CF, Kwiterovich K, Oster-Granite ML, Coyle JT. Lesions of the basal forebrain in newborn mice disrupt cortical cytodifferentiation when visualized by the rapid Golgi method. *Cerebral Cortex* 1991; **1:** 143–157.
59. Kinsman S, Moran T, Reeves R, Gearhart J, Oster-Granite ML. Somatostatin transgenic mice: neurobiologic consequences of neuropeptide gene expression. *Ann Neurol* 1990; **28:** 442.
60. Kinsman S, Moran T, Reeves R, Gearhart J, Oster-Granite ML. Somatostatin transgenic mice: neurobiologic consequences of neuropeptide overexpression. *J Cell Biochem* 1990; **Suppl 14F:** 23.
61. Andrisani OM, Dixon JE. Somatostatin gene regulation *Annu Rev Physiol* 1990; **52:** 793–806.
62. Kinsman *et al.* (submitted).
63. Oster-Granite ML, Höhmann CF, Kinsman SL, Coyle JT. Alterations in levels of choline acetyltransferase in somatostatin transgenic mice. *Soc Neurosci Abstracts* 1991; **17:** 1439.
64. Rubenstein M, Liu B, Goodman RH, Low MJ. Targeted expression of somatostatin in vasopressinergic magnocellular hypothalamic neurons of transgenic mice. *Mol Cell Neurosci* 1992; **3:** 152–161.
65. Breder CD, Yamada Y, Yasua K, Seino S, Saper CB, Bell GI. Differential expression of somatostatin receptor types in brain. *J Neurosci* 1992; **12:** 3920–3934.
66. Plano S, Kinsman SL, Oster-Granite ML, Coyle JT, Hohmann CF. Alterations in somatostatin immunoreactivity in transgenic mice. *Soc Neurosci Abstracts* 1991; **17:** 1439.

Discussion of Presentation

MARY LOU OSTER-GRANITE
(Edited by D.A. Snider and Salvador Castells)

W. T. Brown: Dr. Groner did work with the superoxide dismutase (SOD) gene which is related to the trisomy 21 state. Would you comment?

M. L. Oster-Granite: We have been looking at brain development in Groner's SOD mice. In Down's syndrome, individuals inherit three copies of the SOD gene, and it is believed that some of the problems that these individuals experience occurs because they are unable to handle very toxic products called oxygen radicals. It is thought that having too much SOD leads to increased levels of a very toxic compound called hydrogen peroxide. Groner has studied the phenotypic effects of excess SOD in these transgenics, and he found one phenotypic change that was very typical of Down's syndrome individuals in the mice—an abnormality in the structure of the neuromuscular junctions in the tongue. We have been looking developmentally at the nervous system of these animals; there is not a significant change in pituitary structure or hypothalamic structure that you would anticipate if SOD were the cause of those particular changes you see in Down's syndrome. I favor the notion that it is not a single gene that produces those changes in cell number, function or structure, but rather that it is a constellation of genes influencing each other in development that slowly distorts their function in the trisomic state. His animals do have changes, but they are not as dramatic as those that you see in Down's syndrome, and those that we see in these somatostatin transgenic mice.

G. Merriam: As you are probably aware, there have been other transgenic somatostatin models and in most of them, where they were coupled with companion studies giving growth hormone-releasing hormone (GHRH) or human GH in transgenic studies, they found that human GH or GHRH animals were large, but the somatostatin transgenic animals were usually not small. In your own study you have shown that some of your transgenic lines are growth retarded and others not. I wonder if it is possible that there may be an explanation other than the hormonal effect of somatostatin? Do you know of the insertion site of these transgenes? Is it possible that the genes in the lines where growth is affected have been inserted into the middle of some normal growth gene and have interfered with the function of a normal gene that is important for growth?

M. L. Oster-Granite: That is the reason you make multiple lines of transgenics. We did not use homologous recombination, which would mean that you would be inserting an extra copy right into the endogenous gene to generate these mice. That technique has not been terribly successful in transgenesis. Many of the transgenics that have been made are a sort of composite; you can select a heterologous promotor which makes the selected gene express only under certain conditions. In these transgenics, you can put those conditions which facilitate the promoter's expression into play and see what happens.

G. Merriam: Like a metallothionine?

M. L. Oster-Granite: Yes. These transgenics are made with the endogenous mouse gene 15 kb fragment that we think contains all of the 'beeps, squeaks and whistles', so these animals should be expressing somatostatin in a temporally and spatially appropriate manner during development, and that is exactly what we see in these mice. Now if you do insert the transgene into an ectopic location, you have the opportunity to insert it into a genetically silent region—pieces of the genome that don't do anything. These mice don't express the transgene. You also have the opportunity to insert it into a locus that is another developmental regulator. If that were the case, we would not anticipate, since the insertion is random, that we would be able to create multiple lines which would show a similar phenotype such as growth retardation. We have several lines with growth retardation, but only one with this slightly increased growth effect. Of course, to me, that is the one you want to use for insertional effects because you are looking at a paradoxical effect. So those studies of transgene insertion site mapping must be done. My feeling is that if you get several lines of transgenics you can show a titration effect from very high levels of message and protein expression to relatively low levels. If growth retardation is consonant with an altered level of actual protein that is made, then you are probably looking at a real effect. I tend to feel that is what we are seeing in these animals. It is true, however, that I don't know exactly where the transgene was inserted.

G. Merriam: Is there any evidence for brain lesions, smaller brain size or neuronal deficits in these somatostatin transgenics?

M. L. Oster-Granite: The brains are smaller. We are now doing birth-dating studies on them. Cholinergic markers are affected in line 13.

G. Merriam: I think that if you are postulating it as a neuronal effect versus a hormonal effect, the evidence seems to suggest that this does not parallel Down's syndrome where there is no hypersecretion of somatostatin. In fact, Dr Castells has found that hGH responses to GHRH in Down's syndrome children are at least as brisk as that in control children, and somatostatin being a non-competitive inhibitor of the response to GHRH you see a blunted response to GHRH if ongoing hypersecretion of somatostatin were a major part of the GH deficiency.

M. L. Oster-Granite: You see, in my situation, I have to deal with the fact that trisomy 16 mice have three copies of *preprosomatostatin*, so I have to factor those effects out of the trisomy 16 mouse before I can make some comparisons of what happens in Down's syndrome; that is the reason I generated those mice. I want a baseline of how these animals grow when they have too much somatostatin, compared to how they grow when they have only that little distal cluster of genes in triplicate, so this is a baseline type of information we really need.

Z. Laron: So much somatostatin would probably act also on insulin and glucagon, which has a lot of effects. Would you comment?

M. L. Oster-Granite: We have measured insulin levels in two of our lines. The levels are lower, but they are not significantly lower, and these animals are not overtly diabetic at six months of age. I cannot say what is going to happen to these animals as they get older and older, because we do not have a large enough cohort. We have not yet measured insulin at later ages to see if there is really a decrease in insulin secretion because of an oversecreting delta cell in the pancreas. We know that the transgenics are making more somatostatin in their pancreas, so that each tissue that we look at seems to be overexpressing somatostatin in a tissue-appropriate manner.

Part III

GROWTH HORMONE AND OTHER TREATMENTS IN DOWN'S SYNDROME

11 Evaluation and Treatment of Growth Hormone Deficiency with Growth Hormone-releasing Hormone

GEORGE R. MERRIAM[1,4]* PATRICIA N. PRINZ[3,4] AND ROBERT S. SCHWARTZ[2]

Divisions of [1]Metabolism, Endocrinology and Nutrition and [2]Gerontology, Department of Medicine, and [3]Department of Psychiatry and Behavioral Sciences, University of Washington School of Medicine; [4]Research Service, American Lake VA Medical Center; Seattle, Washington, USA

INTRODUCTION

The secretion of growth hormone (GH) is regulated by at least two hypothalamic peptides which are secreted into the hypothalamic–pituitary portal circulation: somatostatin (SRIH), which inhibits GH secretion; and growth hormone-releasing hormone (GHRH), a 44-amino acid peptide which stimulates GH secretion (Fig. 1). The interaction of these two factors together generates the pattern of episodic GH secretion during the day. While the studies described elsewhere in this volume focus on the therapeutic effects of administering exogenous GH, it is also possible to achieve many of these effects through stimulating the secretion of endogenous GH by administering or manipulating GHRH or SRIH, provided the pituitary somatotrophs are intact. Thus GHRH offers a potential alternative method of treatment which may have certain advantages.

There is still no literature on the effects of GHRH treatment in Down's syndrome, and therefore this chapter will focus on GHRH in 'classical' GH deficiency (GHD). Since all the known effects of GHRH on somatic or linear growth are mediated through its stimulation of GH secretion, however, and since studies of Castells and colleagues have shown that GHRH stimulates a brisk release of GH in short children with Down's syndrome (1), one can with some confidence extrapolate the effects of GHRH to those obtained with GH. This chapter surveys the role played by GHRH in the pathophysiology of GH

*Address for correspondence: Division of Metabolism Endocrinology, and Nutrition, RG-26, University of Washington, Seattle, Washington 98195, USA.

Growth Hormone Treatment in Down's Syndrome. Edited by S. Castells and K.E. Wisniewski

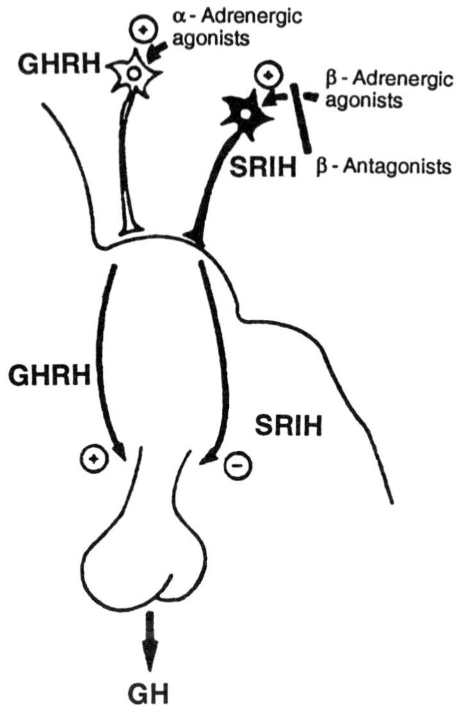

Figure 1. Regulation of GH secretion by GHRH and SRIH, and effect of α- and β-adrenergic neurotransmitters and their blockade on these two hypothalamic peptides. α-Adrenergic agonists stimulate GHRH secretion, while β-agonists stimulate and β-antagonists inhibit SRIH secretion

deficiency and the possible therapeutic role of GHRH in its treatment; the possible utility of neurotransmitter-enhancing adjuvants to potentiate the efficacy of GHRH therapy; the potential of a family of small GH-releasing peptides in the future therapy of GHD; and some suggestions of possible central nervous system (CNS) effects of GHRH.

PATHOPHYSIOLOGY OF GROWTH HORMONE-RELEASING HORMONE IN GROWTH HORMONE DEFICIENCY

Early physiological studies clearly demonstrated that GH secretion is under tonic stimulatory control by the hypothalamus (2). For example, sectioning the pituitary stalk results in low circulating levels of GH. However, the large-scale hypothalamic extraction and fractionation programs of the early 1970s first yielded somatostatin, a 14- and 28-amino acid peptide which inhibits secretion of GH and several other hormones. GHRH itself remained elusive, probably because of its larger size and susceptibility to degradation, and was not isolated until 1982, and then not from hypothalami but from two pancreatic adenomas which secreted GHRH 'ectopically'. In its full sequence it is a 44-amino acid peptide (GHRH(1–44)NH_2), but since the active site is located at the N-terminal, subsequences as short as 29 amino acids retain full biological potency. It is rapidly inactivated by cleavage between positions 2 and 3; most of the immunoassayable material in circulation is thus biologically inactive fragments. GH is not secreted continuously, but rather in pulses, six to eight times or more per day in children and young adults, fewer in older subjects, with greater secretion during slow-wave (delta) sleep. Each secretory episode may reflect the coordinated secretion of GHRH and inhibition of SRIH. Immunoneutralization of GHRH blocks the secretion of GH in animals, supporting the belief that GHRH is the critical physiological stimulator of GH secretion.

Secretion of GHRH and SRIH is also not autonomous, but rather is regulated in turn by a variety of hypothalamic neurotransmitters and neuromodulators, including α-adrenergic agonists, which stimulate GHRH, and β-agonists, which stimulate SRIH and so inhibit GH (Figure 1). In addition to these, dopaminergic and cholinergic agonists also inhibit SRIH secretion. These effects can be used to potentiate GH secretion, as discussed below. A variety of physiological reflexes affect these converging neurotransmitter pathways to produce familiar physiological responses such as the GH responses to slow-wave sleep, stress, exercise, or glucose.

Administered GHRH stimulates a rapid rise in GH secretion to levels which approximate the maximum seen after physiological stimuli: approximately 30–50 μg/l in young adult men and women (3). A maximum response is seen after an intravenous (i.v) dose of approximately 1 μg/kg. The range of responses observed after a single bolus, however, is extremely wide, probably reflecting moment-to-moment variability in endogenous secretion of SRIH, which variably blunts the GH response to the administered GHRH. Since unlike an endogenous burst of secretion, there is usually no coordination between these endogenous changes and the timing of the exogenous bolus, the responses vary widely, even on repeated testing of the same individual (4). Healthy prepubertal children show similarly brisk responses to GHRH, with no difference between boys and girls and little variation across stages of puberty (5).

GH deficiency can have many causes, including aplasia of the pituitary or a deletion in the GH gene. In these circumstances GHRH would be expected to have no effect. However, when children with idiopathic GH deficiency, with responses less than 7 μg/l to at least two standard provocative tests, are tested with GHRH, 70–80% respond with a rise in GH (Fig. 2), indicating that in the majority of cases the pituitary is intrinsically capable of responding and thus the GHD must be due to a deficiency of GHRH. Although the responses overlap into the normal range, they are on average lower than those seen in

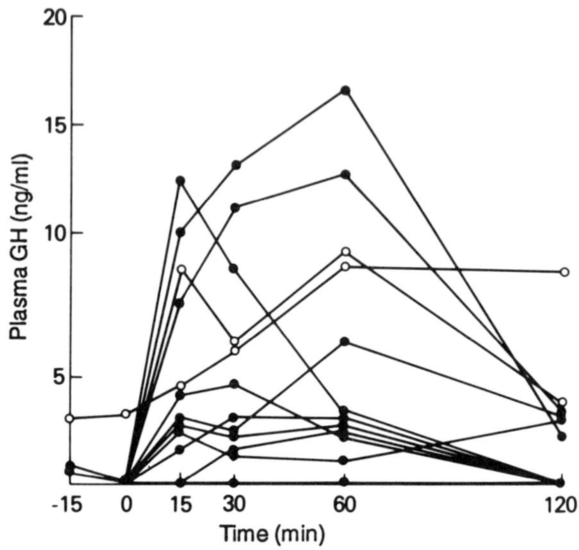

Figure 2. GH responses to a single i.v. bolus of 1 μg/kg GHRH(1–44)NH_2, in children with idiopathic GH deficiency. The majority of subjects show an elevation of GH

normal children; responses in adults with GHD are even lower, and in both GHD children and adults they increase in response to repeated stimulation. These findings are interpreted to reflect atrophy of somatotrophs, with a reduction in GH mRNA and GH synthesis after chronic understimulation by GHRH, and the full or partial reversal of this atrophy on repeated stimulation with GHRH. Aging adult subjects also show a progressive reduction in GH secretion, mediated through a reduction in GHRH and possibly a lesser increase in SRIF secretion, and the possibility that GHRH may have therapeutic value in this large population has also become of wide interest.

The brisk GH response to GHRH in children with reduced or even absent spontaneous GH secretion and low or absent GH responses to other provocative tests, in some cases to many different attempted stimuli, is a remarkable finding. The striking difference in results reflects that conventional diagnostic tests all act indirectly through stimulating endogenous GHRH secretion, while GHRH acts directly on the pituitary to release GH. Thus in the great majority of cases GHD is a hypothalamic disorder resulting in GHRH deficiency, and supplying exogenous GHRH can evoke normal pituitary responses and restore sufficient GH secretion. Castells and colleagues have recently shown that short children with Down's syndrome also respond briskly to GHRH (1).

TREATMENT OF GROWTH HORMONE DEFICIENCY WITH GROWTH HORMONE-RELEASING HORMONE

In patients in whom acute administration of GHRH acutely stimulates GH secretion, repeated administration of GHRH should chronically elevate GH and so might be used to treat GHD. Although the practicality of this approach requires verification, it has theoretical appeal, as it specifically addresses the physiological lesion in GHD—GHRH deficiency—and potentially preserves the feedback relationships between GH secretion and the generation of insulin-like growth factor 1 (IGF-1). There are parallels between this situation and that of patients with hypothalamic amenorrhea due to reduced secretion of hypothalamic gonadotropin-releasing hormone (GnRH): treatment with the pituitary gonadotropins follicle-stimulating hormone (FSH) and luteinizing hormone (LH) bypasses all normal feedback relationships and risks over-stimulating the ovary, while treatment with pulsatile GnRH preserves the normal feedback between ovarian and pituitary hormones and so offers a safer, more self-regulating, and more physiological mode of treatment. Of course, each pituitary regulatory system is different, and suggestions by analogy must be tested directly.

Several studies have now been published which examine the therapeutic effects of treatment with GHRH. In the first of these, we administered GHRH(1–44)NH_2, 1 μg/kg i.v., as repeated bolus doses at 3 h intervals, mimicking the frequency of physiological pulses, over a 2-week treatment period in six GH-deficient children (6). To assess the frequency of responses, GH responses to an acute GHRH dose were not used to select subjects for this trial. Profiles of GH secretion and plasma levels of IGF-1 were obtained. Because this treatment period is too short to demonstrate effects on height, linear growth responses were measured with the lower-leg measuring device of Valk, the sensitivity of which allows assessment of changes in growth velocity over very short treatment periods.

In four of the six subjects, repeated doses of GHRH stimulated pulsatile release of GH (Fig 3). In some cases, the GH response was initially weak, but increased progressively during therapy. GH secretion ceased immediately upon cessation of GHRH therapy. The other two subjects had no initial GH response to GHRH; of these, one began to acquire a measurable response during the treatment period. These increases demonstrate the priming effect of repeated doses in GH-deficient subjects, and most likely reflect an increase in GH synthesis and the size of the releasable GH pool. The remaining one subject had no response to GHRH even after repeated stimulation.

Plasma levels of IGF-1 rose in most subjects, but were unaffected in the GH-unresponsive patient. Four of the six patients showed an increase in lower-leg growth velocity; the GH-unresponsive patient, and also the one acquiring a GH response only late in the short course of treatment, showed no effect.

While this early study clearly demonstrates the efficacy of GHRH therapy

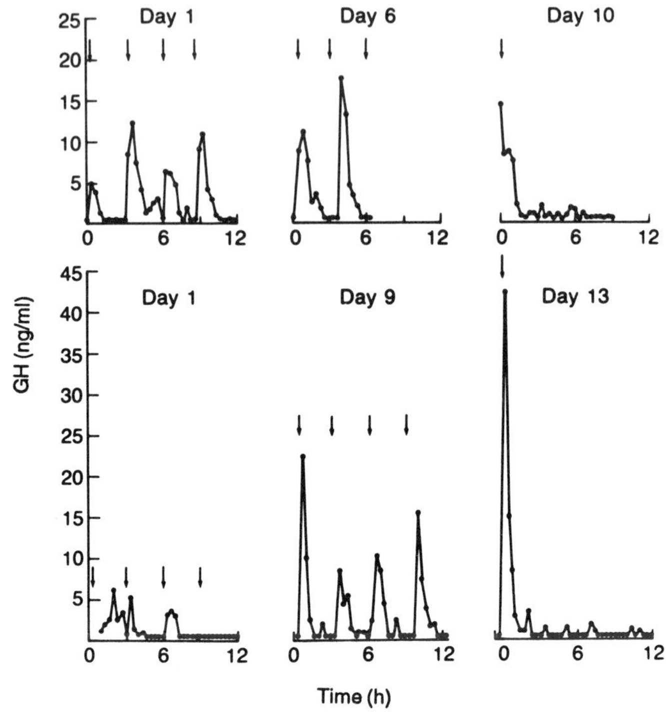

Figure 3. GH responses to repeated i.v. bolus doses of GHRH, 1 μg/kg per dose in children with idiopathic GH deficiency. From Gelato *et al.* (6)

in a majority of even unselected GHD patients, i.v. therapy is obviously impractical for chronic therapy, and administering GHRH as pulses even with the portable miniature pumps used for GnRH may not be practicable in children. Subsequent studies, however, have shown that subcutaneous (s.c.) therapy with somewhat higher doses of GHRH can also be effective, and even that a pulsatile mode of administration may not be necessary. Two studies illuminate this point. Malozowski and colleagues administered GHRH in a variety of dosage schedules designed to test effects of both the frequency and the total daily dose of GHRH—as s.c. injections of 1, 3 or 9 μg/kg daily or three times daily; and by pump in boluses of 3 μg/kg 1–17 times daily (7). Both by pump and as discrete injections, GHRH was capable of restoring linear growth velocity when given in doses exceeding 9 μg/kg per day, regardless of the schedule; thus 9 μg/kg once daily was as effective as 3 μg/kg three times daily. Furthermore, the pattern of overnight GH secretion was similar, and showed similar numbers of GH pulses even when the pump frequency was changed. Vance and colleagues showed that even a continuous infusion of GHRH generated a pulsatile, not a continuous pattern of GH release, with day–night differences mimicking the normal physiological pattern (8). The explanation for this effect, as well as for the relative pattern insensitivity in GHRH therapy, is most likely variations in endogenous SRIH secretion occurring even in GH-deficient subjects, variably modulating the effect of even a continuous infusion of GHRH. This not only makes for a greatly simplified and thus far more practicable dosing schedule but may make another argument for the physiological superiority of GHRH therapy, since there is some evidence with GH therapy that a pulsatile mode of administration may be more effective than continuous administration or larger doses at infrequent intervals (9, 10).

Several studies have now documented the effects of long-term s.c. therapy with GHRH. Rochiccioli and colleagues administered 10μg/kg GHRH(1–44)NH_2 as a single evening dose for 12 months in six GHD subjects (11). All showed a brisk initial acceleration of growth. However, in most there was a decline in this response over time, as may be seen in GH treatment as well.

Thus although the number of subjects treated with GHRH worldwide is still relatively small and there are some non-responders requiring treatment with GH, this form of treatment still shows promise. It is generally effective in accelerating growth velocity and restores a pulsatile pattern of GH secretion, quite different from the unphysiological pattern of GH levels after GH administration. A regimen of daily s.c. injections is the same as one of the dosage schedules used for GH therapy; and in principle a 29-amino acid peptide using the pituitary as a biological amplifier should be less expensive to prepare than synthetic 192-amino acid GH. Its rapid inactivation in serum and blunting of its effects by SRIH pose the major limitations on its potency. A GHRH superagonist with longer duration of action and possible increased potency would clearly be desirable and, unlike GnRH analogs, would likely continue to stimulate pituitary secretion, allowing for a further reduction in dose frequency while maintaining pulsatile GH secretion. However, this has been an elusive goal. Despite extensive study of position-2 modified analogs of GHRH with promising properties in animals, no GHRH analog with superagonist properties in man has so far been devised. Work on slow-release formulations or electrophoretic skin patch devices providing a prolonged action with native GHRH is progressing but has not yet reached clinical utility.

NEUROTRANSMITTER ENHANCEMENT OF THE EFFECTS OF GROWTH HORMONE-RELEASING HORMONE

Another potential way to enhance the effects of GHRH would be to inhibit the actions of SRIH, which intermittently and usually unpredictably blocks GHRH stimulation of GH release. Since this inhibition is non-competitive it cannot be overcome even by increasing the dose of GHRH. As noted above,

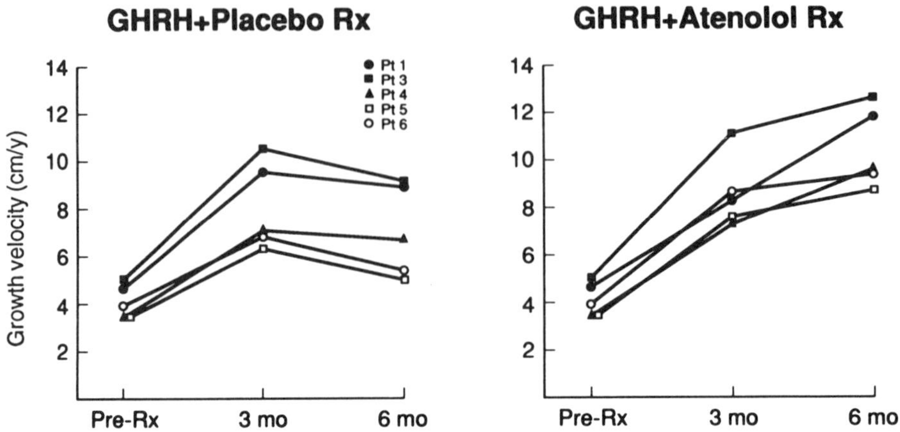

Figure 4. Changes in growth velocity during treatment with GHRH plus placebo versus with GHRH plus the β-antagonist atenolol. These preliminary results suggest an enhancement by atenolol, but should be regarded as tentative pending the outcome of a larger trial. From Cristiano *et al.* (14)

β-adrenergic agonists stimulate SRIH secretion (Fig. 1), and thus β-adrenergic antagonists might reduce SRIH secretion.

This is apparently the case. Pretreatment with the β-antagonist propranolol significantly increases the GH response to GHRH. A similar enhancement has been reported for the selective β-antagonists metoprolol and atenolol (12), and also with dopaminergic agonists such as bromocriptine and stimulators of cholinergic activity such as pyridostigmine.

For these effects to be useful in enhancing the therapeutic effect of GHRH, the augmentation of GH release would need to persist chronically; and with some agents such as pyridostigmine this has been difficult to demonstrate (13). We are conducting a collaborative study of the ability of atenolol co-administration to enhance the growth-stimulating effects of chronic therapy with GHRH. Results of this trial are still pending, but a preliminary study suggests that GH-deficient children receiving GHRH plus atenolol may have a greater increase in growth velocity than those receiving the same dose of GHRH plus placebo (Fig. 4) (14). Such an approach bears further study and might prove particularly useful in treating patients with increased adiposity, quite common in Down's syndrome, since the reduced GH secretion in obesity appears to be related to increased secretion of SRIH.

GROWTH HORMONE-RELEASING PEPTIDES

While GHRH is the major stimulator of physiological GH release, another and completely different class of peptides has also been found to be a potent stimulator of GH secretion. These GH-releasing peptides (GHRPs) were synthesized by Bowers and Folkers as successively modified derivatives of enkephalins, which also have GH-releasing effects (15). They are small six to seven amino acid peptides with D-amino acids in their primary structure. They do not bind to the GHRH receptor, and their dose–response characteristics for release of GH are not parallel to that of GHRH, producing a greater release of GH than even a maximally stimulating dose of GHRH (Fig. 5) (16). In addition, the action of GHRP appears greater *in vivo* than *in vitro* and can synergize with co-administered GHRH.

These unusual findings admit of several possible explanations, including the possibility that their actions synergize directly on the pituitary; but also that GHRP acts to stimulate GH not only by direct pituitary action, but also by indirect action via the hypothalamus; hence the greater potency *in vivo*. One possible way that GHRP could do this would be to inhibit SRIH secretion, allowing a greater maximal GH response. However, two lines of evidence suggest that this is not the case. For one, β-adrenergic antagonists enhance the effects of GHRP to the same degree as they do the effects of GHRH (Fig. 6) (16), which would not be expected if SRIH were already suppressed. In

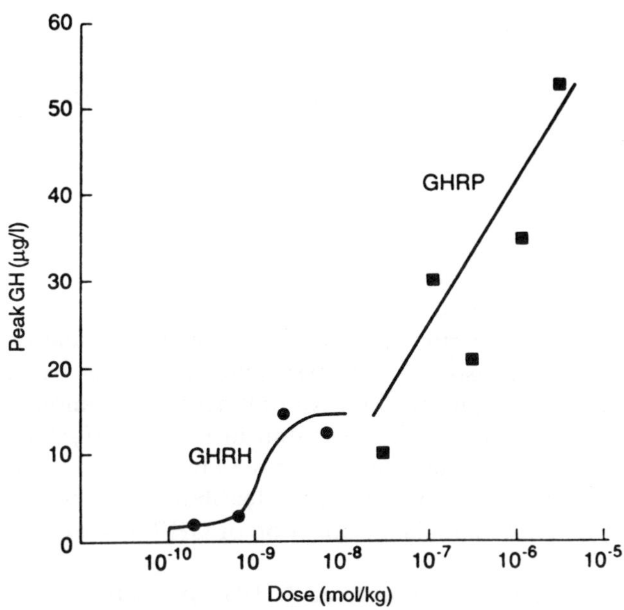

Figure 5. Dose–response curves for the stimulation of GH secretion by i.v. bolus doses of GHRH(1–44)NH_2 (•) and by GHRP (■) in tethered awake monkeys *in vivo*. The two curves are markedly dissimilar; higher molar doses of GHRP are required to stimulate GH secretion, but higher levels of GH are produced, with no evidence for maximal effects having been reached. From Malozowski *et al.* (16)

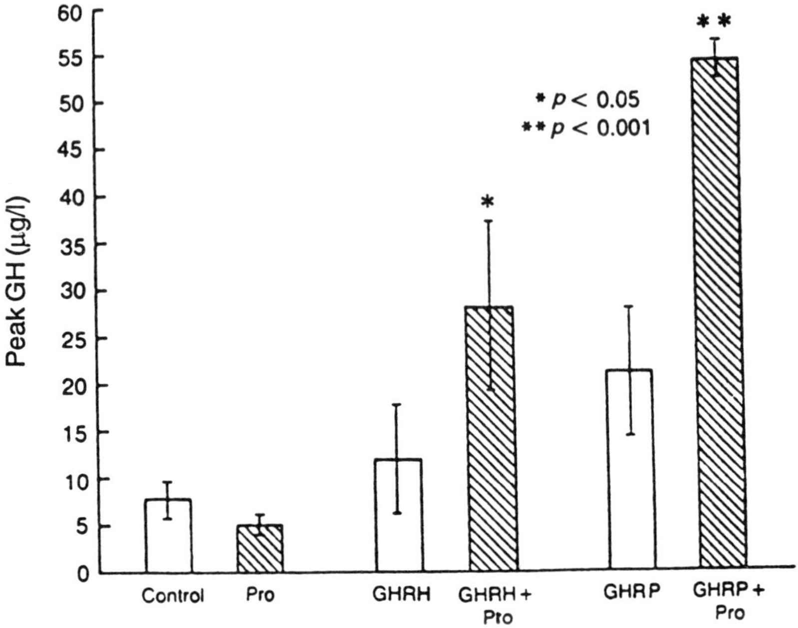

Figure 6. Effect of propranolol pretreatment on the GH responses to GHRH and GHRP in tethered awake macaques. Propranolol (Pro) had no effect on spontaneous GH secretion (left pair of bars), but approximately doubled the peak GH responses to both GHRH (center bars) and GHRP (right bars). From Malozowski *et al.* (16)

addition, GHRP administered to hypothalami incubated in organ culture releases rather than suppresses SRIH in high doses (17).

In principle GHRPs could be extremely attractive agents for the therapy of GH deficiency. Their D-amino acids resist proteolytic degradation; their small size makes them inexpensive to synthesize and suitable for transnasal, transdermal, or even oral administration. An oral agent to promote GH secretion could have enormous practical application in therapy. However, the possibility that hypothalamic reflexes might form an important part of their mechanism of action raises questions about the feasibility of using GHRP to treat GH deficiency, since hypothalamic secretion of GHRH is impaired in most of these patients.

To study this issue, we administered GHRP to children with documented GH deficiency, and compared the response to that to a bolus of GHRH (18). A single bolus of GHRP was nearly as effective as one of GHRH in these patients (Fig. 7). Because of the different dose–response curves and the lack of priming it is not possible to determine its potency relative to GHRH, but it provoked a significant GH response in almost as large a fraction of the

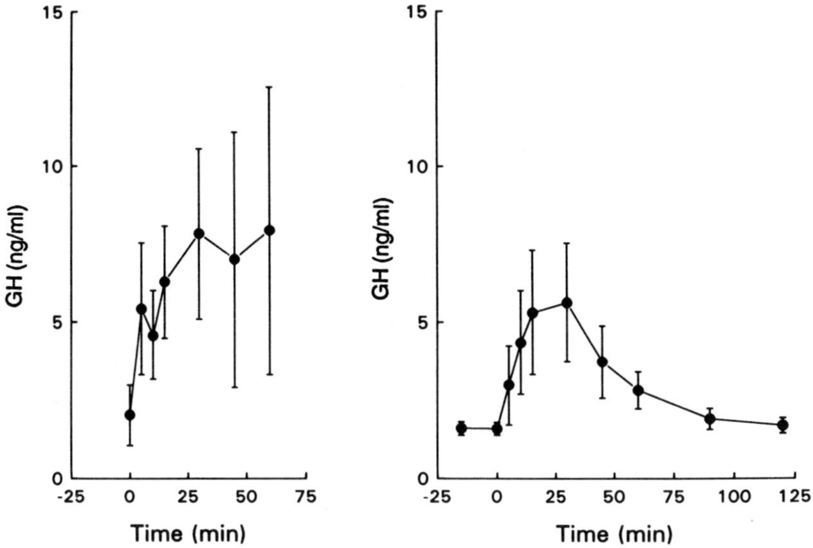

Figure 7. Effects of bolus doses of GHRH (left) and GHRP (right) on GH secretion in 13 GH-deficient children. Sampling was terminated 60 minutes after the GHRH injections, and thus the declining limb of these curves is not shown. From Mericq *et al.* (18)

subjects. Thus GHRPs remain of great interest as a possible alternative therapy. A first trial of therapy of GHD with GHRP is now underway.

CENTRAL NERVOUS SYSTEM EFFECTS OF GROWTH HORMONE-RELEASING HORMONE

The treatment of Down's syndrome children with GH or GHRH initially focused on the acceleration of somatic growth to increase final height. There has been a surge of further interest based on suggestions that treatment might also improve neurological function in these children. There is still very little scientific information on this subject.

The most extensive information on the CNS effects of GH and GHRH derives from studies of sleep. Although sleep and higher cognitive function are sharply distinct, sleep studies do have several advantages as a CNS measure. There are standardized sleep laboratory methodologies which can be widely reproduced, and one can examine both acute and chronic treatment effects on sleep patterns.

Endogenous GH secretion and the response to administered GHRH are enhanced during slow-wave sleep (SWS), most likely reflecting an effect of sleep stages on SRIH and GHRH release. But can there be effects of GHRH *upon* sleep? Åstrom and colleagues report an abnormality of sleep patterns in patients with GH deficiency, with a significant decrease in slow-wave (delta) sleep (19, 20). Recent preliminary studies report that a single bolus of GHRH can acutely enhance SWS over 1–2 h, while somatostatin had no effect upon SWS (21, 22). Whether the effect was directly due to GHRH or to the GH it stimulated is not clear from these short-term experiments; the rapidity of the effect makes it unlikely that an increase in IGF-1 is responsible. One study reports that GH treatment decreases total sleep but increases rapid eye movement (REM) sleep (23). This result from chronic therapy is not directly comparable with those observed in the very short-term GHRH studies but leaves open the possibility of an effect of GHRH which is distinct from that of GH. Further study of these sleep effects and their mechanisms may provide a useful basis for designing the studies of the chronic and cognitive effects of GH and GHRH which have not yet been performed.

SUMMARY

Since most cases of GH deficiency are due to a deficiency of hypothalamic GHRH, GHRH offers an alternative approach to the treatment of GHD which is more physiological and can be highly effective. Newer synthetic GHRPs work through a different mechanism to stimulate GH secretion and may in the future offer an orally active therapy of GHD. These GHRPs synergize with GHRH in releasing GH; the efficacy of other agents to enhance the therapeutic effects of GHRH is under study. There appear to be some acute effects of GHRH or GH on sleep EEG rhythms; the question of whether these effects point to more general neurological effects of GH is still an open one.

ACKNOWLEDGEMENTS

Many colleagues, students and fellows at the National Institutes of Health and collaborators at the University of Washington, Tulane University and the Universidad de Chile contributed to the studies summarized here, particularly Dr Marie Gelato, and also Drs Cy Bowers, Fernando Cassorla, Audrey Cristiano, Enhui Hao, D. Lynn Loriaux, Saul Malozowski, Veronica Mericq and Ora Pescovitz. These studies would not have been possible without the active collaboration of the Clinical Center professional nursing staff. We are grateful for the technical assistance of Stephen Galt. This work was supported in part by the National Institutes of Health (R01 AG10943) and the Veterans Health Services and Research Administration.

REFERENCES

1. Castells S, Beaulieu I, Torrado C, Wisniewski KE, Zarny S, Gelato MC. Hypothalamic versus pituitary dysfunction in Down's syndrome as a cause for growth retardation. International Conference on GH Therapy in Down's Syndrome, New York, 10–11 October 1992 (abstract).
2. Gelato MC, Merriam GR. Growth hormone-releasing hormone. *Annu Rev Physiol* 1986; **48:** 569–591.
3. Gelato MC, Pescovitz OH, Cassorla F, Loriaux DL, Merriam GR. Dose–response relationships for the effects of growth hormone-releasing hormone(1–44)NH_2 in men and women. *J Clin Endocrinol Metab* 1984; **59:** 197–201.
4. Fornito MC, Calogero AE, Mongioi A, Coniglione F, Vicari E, Moncada ML, D'Agata R, Merriam GR. Intra- and inter-individual variability in growth hormone responses to GH-releasing hormone. *J Neuroendocrinol* 1990; **2:** 87–90.
5. Gelato MC, Malozowski S, Nicoletti MD, Ross JL, Pescovitz OH, Rose SR, Loriaux DL, Cassorla F, Merriam GR. Growth hormone responses to growth-hormone releasing hormone during pubertal development in normal boys and girls: comparison to idiopathic short stature and GH deficiency. *J Clin Endocrinol Metab* 1986; **63:** 173–179.
6. Gelato MC, Rittmaster R, Pescovitz OH, D'Agata R, Nixon W, Loriaux DL, Merriam GR. Effects of pulsatile administration of growth hormone-releasing hormone on short-term linear growth in children with growth hormone deficiency. *J Clin Endocrinol Metab* 1985; **61:** 444–450.
7. Malozowski S, Caruso-Nicoletti M, Rose S, Hench K, Muir-Nash J, Cassorla F, Merriam GR, Gelato MC. GH-releasing hormone (GHRH) therapy of children with GH deficiency: a comparison of infusion pump and manual GHRH treatment regimens. *Endocrinology* 1986; **118:** 216A.
8. Vance ML, Kaiser DL, Evans WS, Furlanetto R, Vale W, Rivier J, Thorner MO. Pulsatile growth hormone secretion in normal men during continuous 24-hour infusion of human growth hormone releasing factor (1–40): evidence for intermittent somatostatin secretion. *J. Clin Invest* 1985; **75:** 1584–1590.
9. Jansson JO, Albertsson-Wikland K, Eden S, Thorngren KG, Isaksson O. Circumstantial evidence for a role of the secretory pattern of GH in the control of body growth. *Acta Endocrinol* 1982; **99:** 24–30.
10. Clark RG, Jansson JO, Isaksson O, Robinson ICAF. Intravenous GH: growth responses to patterned infusions. *J Endocrinol* 1985; **104:** 53–61.
11. Rochiccioli PE, Tauber MT, Coude FX, Arnone M, Morre M, Uboldi F, Barbeau C. Results of 1-year growth hormone (GH)-releasing hormone-(1–44) treatment on growth, somatomedin-C, and 24-hour GH secretion in six children with partial GH deficiency. *J Clin Endocrinol Metab* 1987; **65:** 268–274.
12. Martha PM Jr, Blizzard RM, Rogol AD. Atenolol enhances GH release to exogenous GHRH. *Pediatr Res* 1988; **23:** 393–397.
13. Kirk JM, Ross RJ, Trainer PJ, Froud AL, Davies SC, Savage MO, Besser GM. Pyridostigmine fails to increase either spontaneous or GHRH-stimulated GH secretion during day or night in growth hormone-insufficient children. *Clin Endocrinol* 1991; **34:** 407–411.
14. Cristiano AM, Hao EH, Rose SR, Municchi G, Ma N, Cassorla F, Merriam GR. Co-treatment with a beta-adrenergic blocking agent can enhance the growth response to GHRH therapy in GH-deficient children. Endocrine Society 1991 annual meeting, Washington, DC, abstract 1573.
15. Bowers CY, Sartor AO, Reynolds GA, Badger TM. On the actions of the growth hormone-releasing peptide, GHRP. *Endocrinology* 1991; **128:** 2027–2035.

16. Malozowski S, Hao EH, Ren SG, Marin G, Liu L, Southers J, Merriam GR. GH responses to the hexapeptide GH-releasing peptide and GH-releasing hormone in cynomolgus macaques: evidence for non-GH mediated responses. *J Clin Endocrinol Metab* 1991; **73:** 314–317.
17. Hao EH, Malozowski S, Ren SG, Marin G, Southers J, Merriam GR. A comparison of the effects of GH-releasing hormone and GH-releasing peptide on GH and somatostatin release. *Endocrinology* 1988; **122:** 123A.
18. Mericq V, Cassorla F, Garcia H, Avila A, Merriam GR. Growth hormone responses to a second-generation GH-releasing peptide and to GH-releasing hormone (GHRH) in growth-hormone deficient children. Endocrine Society 1992 annual meeting, San Antonio, TX, abstract 477.
19. Åstrom C, Jochumsen PL. Decrease in delta sleep in GH deficiency assessed by a new power spectrum analysis. *Sleep* 1989; **12:** 508–515.
20. Åstrom C, Lindholm J. GH deficient young adults have decreased deep sleep. *Neuroendocrinology* 1990; **51:** 82–84.
21. Kerkhofs M, Van Cauter E, Van Onderberghen A, Caufriez A, Thorner MO, Copinschi G. Growth hormone releasing hormone (GHRH) has sleep-promoting effects in man. *J Sleep Res* 1992; **1:** 115.
22. Steiger A, Guldner J, Hemmeter U, Rothe B, Wiedemann K, Holsboer F. Effects of growth hormone-releasing hormone and somatostatin on sleep EEG and nocturnal secretion cortisol and growth hormone in man. *J Sleep Res* 1992; **1:** 220.
23. Åstrom C, Pedersen SA. Lindholm J. The influence of GH on sleep in adults with GH deficiency. *Clin Endocrinol* 1990; **33:** 495–500.

12 Growth Hormone, Insulin-like Growth Factor 1 and Brain Growth and Function

Z. LARON,* L. LAZAR AND B. KLINGER
Institute of Pediatric and Adolescent Endocrinology, Children's Medical Center of Israel, Petah Tikva, and the Sackler Faculty of Medicine, Tel Aviv University, Israel

In early fetal development, the brain has a preferential position. Even in the newborn state, the weight of the brain represents 12% of body weight, compared to 2% in adults. What is the role of growth hormone (GH) or its mediator, insulin-like growth factor 1 (IGF-1), (somatomedin-C) in brain development?

Despite the large amount of literature written on the effects of human growth hormone (hGH) on linear growth in children with GH deficiency (1), or with short stature and normal GH secretion (2, 3), there is little information on the effect of GH on head growth and brain development, and even less on the role of IGF-1, the effector hormone of GH (4).

We had the opportunity to gain information on this issue during the investigation of two unique models of GH inactivity caused by hereditary molecular defects. One is hGH gene deletion (5) and the second is a molecular defect of the GH receptor (6), so-called Laron's syndrome (LS—also known as Laron-type dwarfism—LTD) (7–9).

In both instances the fetus and later, the infant, generate little IGF-1—in the first disease in the absence of cumulative hGH and in the second disease in the presence of high amounts of hGH. We studied the consequences of hGH and IGF-1 deprivation and subsequently the influence of hGH or IGF-1 administration on cranial growth.

Intrauterine hGH or IGF-1 deficiency leads to below normal head circumference in most instances, in addition to linear growth retardation (Table 1, Figs 1 and 2). LS patients who had until recently an untreatable disease, continued to have subnormal head circumference throughout childhood and puberty (Figs 3 and 4).

*Zvi Laron is Incumbent of the Irene and Nicholas Marsh Chair in Endocrinology and Juvenile Diabetes, Sackler Faculty of Medicine, Tel Aviv University. Address for correspondence: Institute of Pediatric and Adolescent Endocrinology, Children's Medical Center of Israel, Beilinson Campus, Petah Tikva 49 100, Israel.

Growth Hormone Treatment in Down's Syndrome. Edited by S. Castells and K.E. Wisniewski

Table 1. Head circumference and height in early childhood in six children with hGH gene deletion (expressed as deviation from the 50th centile)

Patient		Age (years)		Head circumference	Height
No.	Sex	CA	BA	(SD)	(SDS)
1	F	0.8	0.1	−2.0	−4.0
2	M	0.8	0.1	−2.2	−5.5
3	F	1.1	0.2	−4.8	−4.0
4	M	2.4	0.7	−3.5	−6.8
5	F	3.2	0.7	−2.4	−7.6
6	M	4.4	1.5	−2.8	−6.7
mean		3.2	0.5	−2.9	−5.8
±SD		1.1	0.5	1.0	1.4

SD = standard deviation; SDS = standard deviation score; CA = chronological age; BA = bone (skeletal) age.

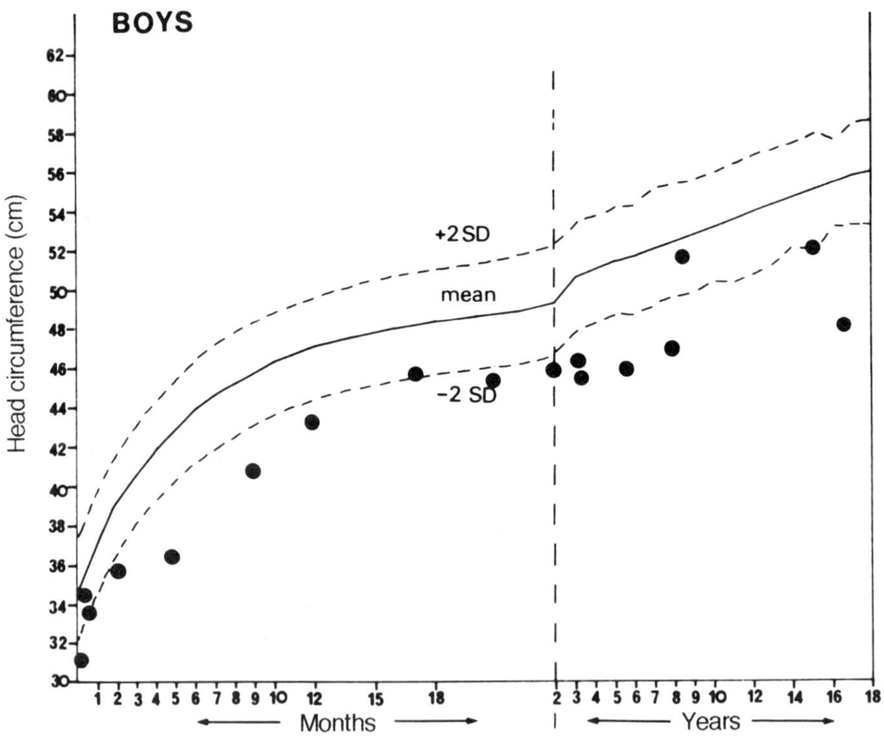

Figure 1. Head circumference in 17 male patients with Laron's syndrome at their first examination (norms according to Nellhaus charts (10))

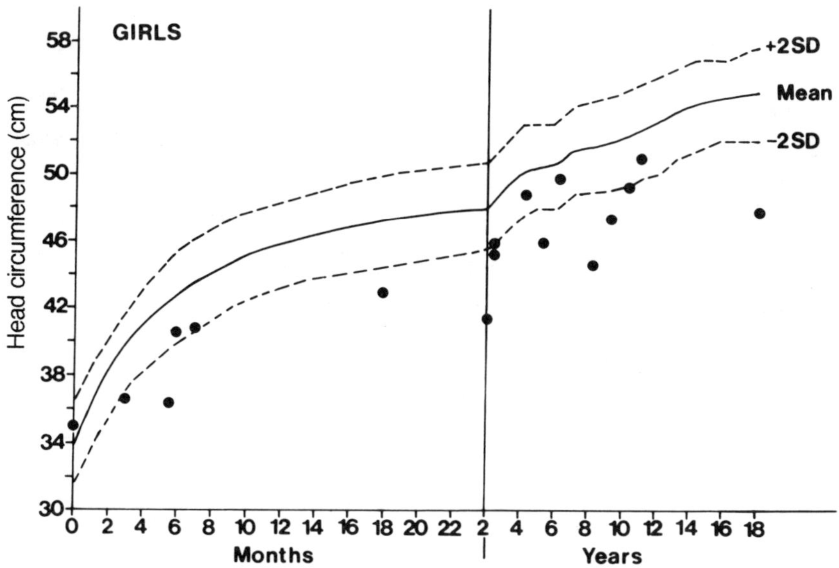

Figure 2. Head circumference in 17 female patients with Laron's syndrome at their first examination (norms according to Nellhaus charts (10))

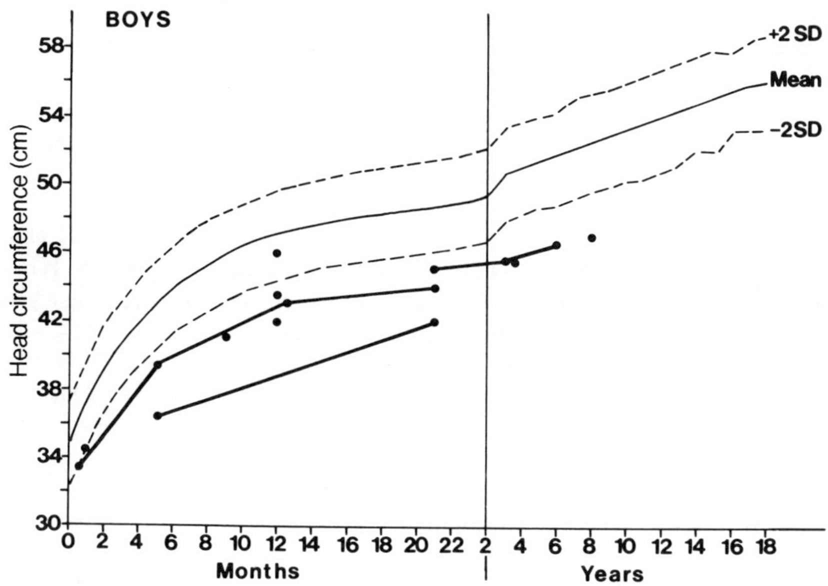

Figure 3. Longitudinal growth of head circumference in three out of nine boys with Laron's syndrome

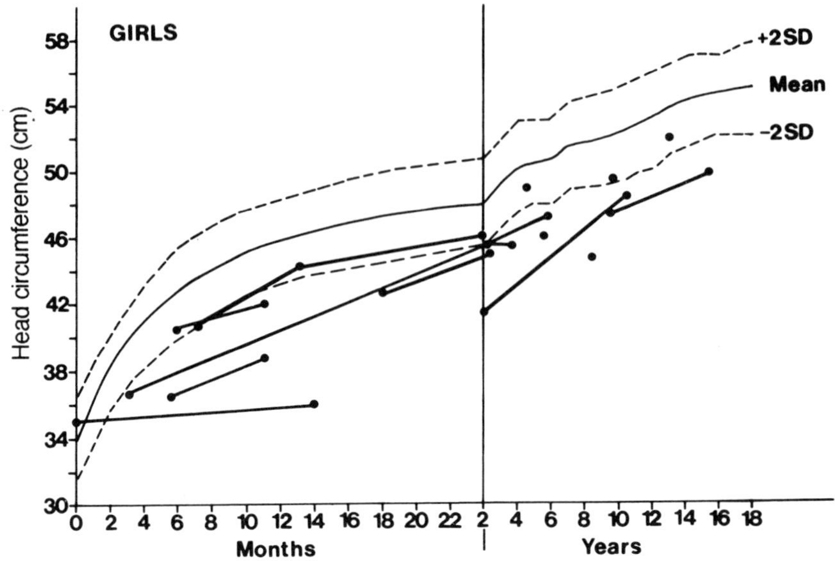

Figure 4. Longitudinal growth of head circumference in eight out of fourteen girls with Laron's syndrome

In previous (11) and recent studies (12) we showed that GH treatment in infants and young children with hGH gene deletion not only accelerates their linear growth but also intensifies the growth of the head circumference, which normalizes faster than the short stature (Fig. 5).

We also found that GH-deficient children of etiologies other than gene deletion have small head circumference which can be normalized if treated before age four years (12). GH treatment started in older children has no effect on head circumference, despite accelerating linear growth.

The biosynthesis of IGF-1 by recombinant DNA technique (13) enabled clinical use of this hormone. Daily subcutaneous administration of IGF-1 to children with LS revealed not only marked enhancement of the linear growth but also a striking growth in head circumference (Figs 6 and 7) (14, 15). Of note is the observation that not only very young children but also those beyond age 10 years showed a significant growth acceleration in their head circumference (Fig. 7).

DISCUSSION

Our studies (11, 12) as well as observations by others (16, 17) show that in GH deficiency starting *in utero* or early infancy there is an under development not only of the chondrocranium, i.e. facial bones of enchondral origin (7, 8), but also of the neurocranium, i.e. the membranous skull known to be dependent on brain growth (18). This GH-dependent effect was observed to occur only in infants and children below age four years who, due to their retarded bone age, can also be considered infants. It thus seems that there is an age or developmental limiting factor to the GH effect on head circumference.

As GH has only a few independent non-IGF-1-mediated effects (19, 20), the question arises whether the observed acceleration of cranial growth is not an IGF-1 effect. Indeed, the demonstration by our group that long-term IGF-1

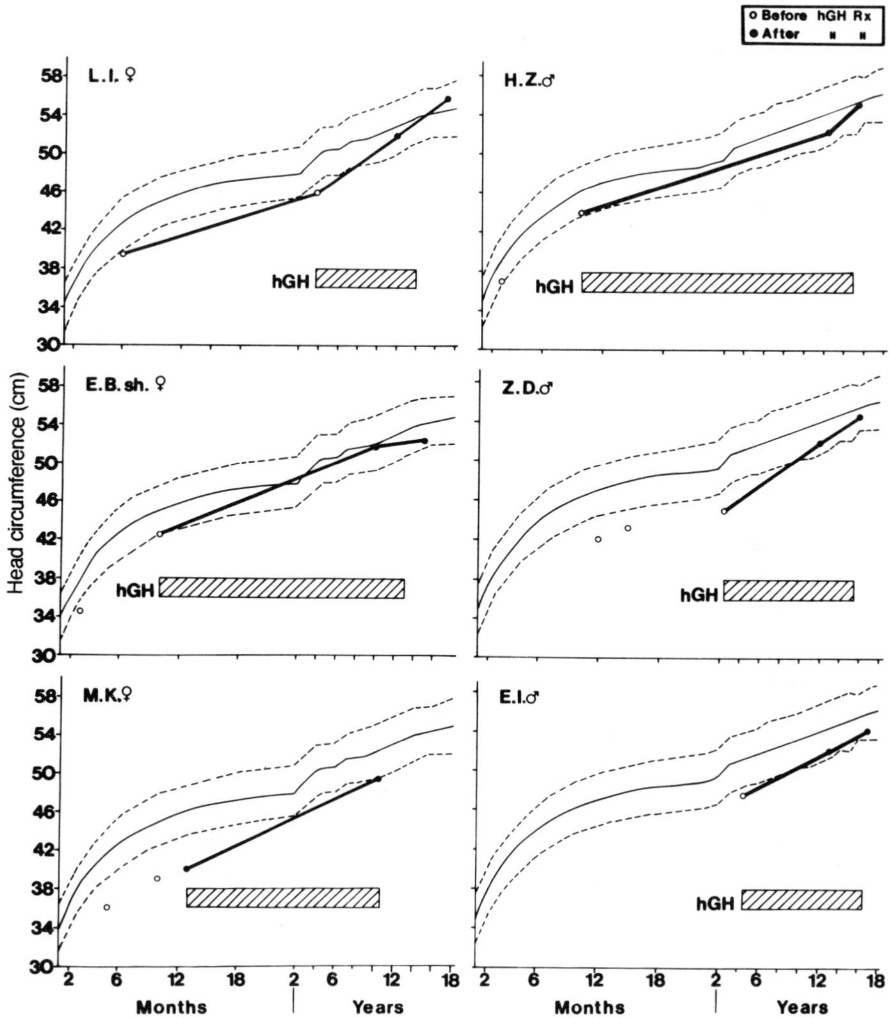

Figure 5. Effect of growth hormone (hGH) treatment on head circumference in six children with isolated growth hormone deficiency due to hGH gene deletion

treatment of children with LS induces a rapid and striking acceleration of cranial growth in addition to increasing linear growth velocity (14) demonstrates that the GH exerts its effect via IGF-1.

Roentgenological and computer tomography studies of some of the patients revealed that the increase in head circumference was not due to hydrocephalus or thickening of the cranial bones. To explain the clinical observations, one needs evidence that IGF-1 influences brain development. This evidence exists: IGF-like immunoreactive material and IGF mRNA have been identified in

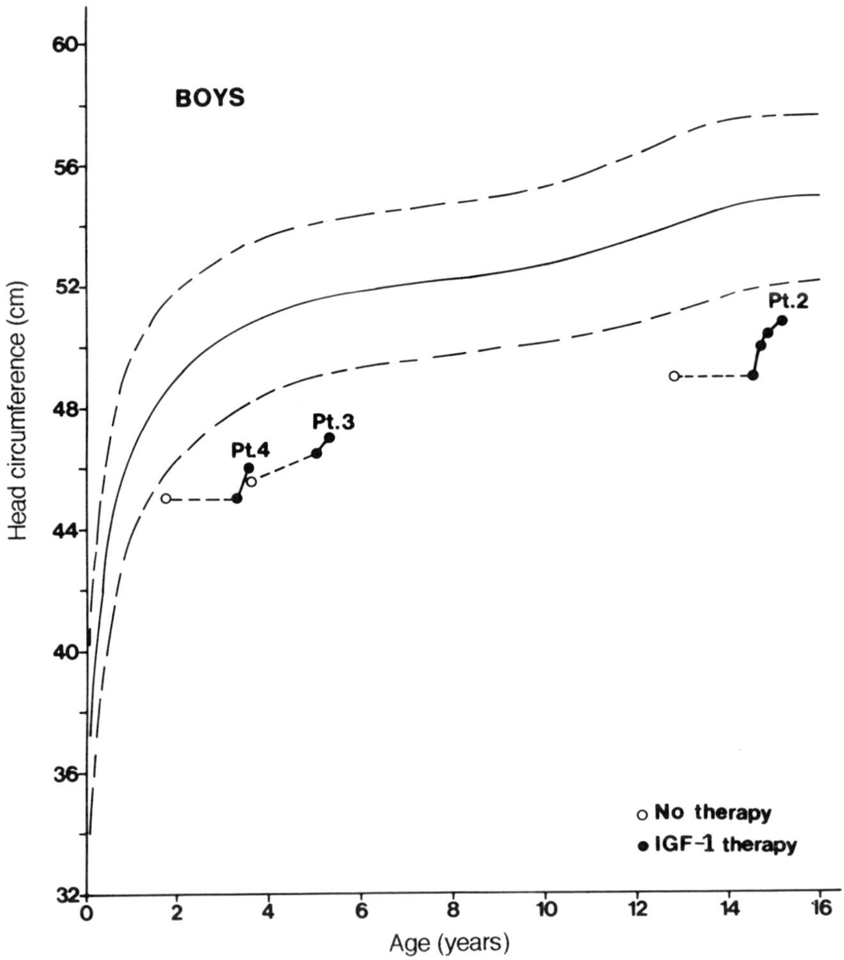

Figure 6. Acceleration of head circumference growth by IGF-1 treatment (120 μg/kg per day s.c.) in three boys with Laron's syndrome. Reproduced with permission from Laron *et al.* (14)

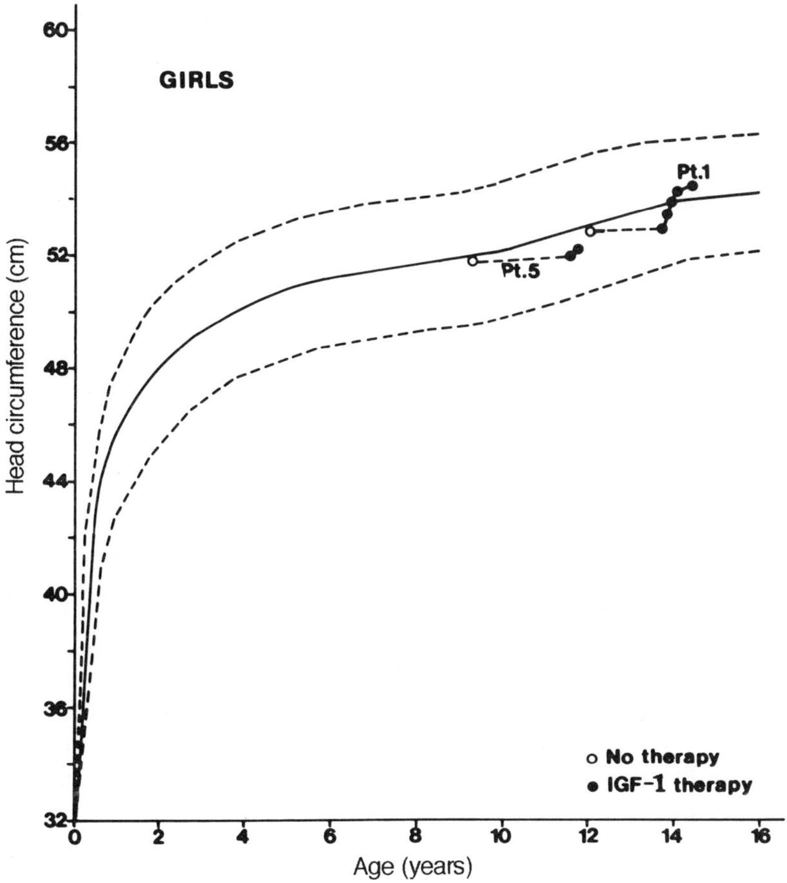

Figure 7. Acceleration of head circumference growth by IGF-1 treatment (120 μg/kg per day s.c.) in two girls with Laron's syndrome. Reproduced with permission from Laron *et al.* (14)

animal (21, 22) as well as human brain tissue (23, 24). The number of IGF-1 receptors seems to be greater in fetal life than postnatally. It has been suggested by Sara *et al.* (25) that a truncated amino-terminal tripeptide gly-pro-glu of IGF-1 is biologically active in the brain as a neuroactive peptide with a potent stimulating action on acetylcholine release. IGF-1 has been shown to stimulate DNA synthesis in fetal brain cells (26) and to induce oligodendrocyte differentiation (27). IGF-2 which has none of the anabolic actions of IGF-1 (4), is considered a major factor in brain development, and among many still unknown effects has been linked to neurite formation (28).

The role of IGF-1, a potent mitogenic and metabolic factor in the development of the nervous tissue seems thus to be documented although in a patchy manner. If indeed IGF-1 affects nervous tissue development, an influence on the functional development of the brain during critical pre- and postnatal periods is expected. Indeed psychological studies in our clinic have shown that patients with isolated growth hormone deficiency (IGHD) and LS have certain intellectual defects (29–31). Despite the fact that early childhood hypoglycemia, which characterizes GH and IGF-1 deficiency (7–9, 32), may contribute to the central nervous system damage, the difference in intellectual capacity observed between GH deficiency which was treated soon after diagnosis and LS patients who until recently were untreated may point to a direct IGF-1 effect as a contributing, if not major cause of the intellectual deficit.

In summary, the long-term auxological study of a unique group of children with pre- and perinatal primary or secondary IGF-1 deficiency, such as patients with LS and hGH gene deletion, revealed that intrauterine and perinatal IGF-1 deprivation led to underdevelopment of the brain and thus to a small cranium.

Early institution of replacement therapy with hGH or IGF-1 induces a rapid acceleration of the brain growth as long as the bony sutures seem open. As IGF-1 seems important also for brain function, the effectiveness of IGF-1 (or hGH) replacement may be related to a critical time period in early life. Therefore it seems imperative to diagnose and treat primary or secondary IGF-1 deficiency early. Now that clinical grade IGF-1 is available as well as unlimited amounts of biosynthetic hGH, no child in need of either of these two hormones should remain untreated. Unfortunately, there are still countries where even diagnosed patients remain untreated, leading to brain deficits in addition to dwarfism.

REFERENCES

1. Laron Z, Butenandt O. *Clinical Use of Growth Hormone: Present and Future Aspects*. Pediatric and Adolescent Endocrinology, Vol. 16, Karger, Basel, 1987; pp. 1–284.
2. Laron Z. Somatomedin-I in clinical use: facts and potential. *Drugs* 1993; **45:** 1–8.
3. Hindmarsh PC, Bridges NA, Brook CGD. Wider indications for treatment with biosynthetic human growth to hormone in children. *Clin Endocrinol* 1991; **34:** 417–427.
4. Humburg RE. Insulin-like growth factor I and II. *Eur J Biochem* 1990; **190:** 445–462.
5. Laron Z, Kelijman M, Pertzelan A, Keret R, Shoffner JM, Parks JS. Human growth hormone gene deletion without antibody formation of growth arrest during treatment: a new disease entity? *Israel J Med Sci* 1985; **21:** 999–1006.
6. Eshet R, Laron Z, Pertzelan A, Dintzman M. Defect of human growth hormone receptors in the liver of two patients with Laron type dwarfism. *Israel J Med Sci* 1984; **20:** 8–11.
7. Laron Z, Pertzelan A, Mannheimer S. Genetic pituitary dwarfism with high serum

concentration of growth hormone: a new inborn error of metabolism? *Israel J Med Sci* 1966; **2:** 152–155.
8. Laron Z, Pertzelan A, Karp M. Pituitary dwarfism with high serum levels of growth hormone. *Israel J Med Sci* 1968; **4:** 883–894.
9. Laron Z. Laron type dwarfism (hereditary somatomedin deficiency): a review. In *Advances in Internal Medicine and Pediatrics*, edited by von Frick P, von Harnack GA, Kochsiek K, Martini GA, Prader A, Springer-Verlag, Berlin 1984; pp. 117–150.
10. Nellhaus G. Head circumference from birth to eighteen years. *Pediatrics* 1967; **41:** 106–114.
11. Laron Z, Roitman A, Kauli R. Effect of human growth hormone therapy on head circumference in children with hypopituitarism. *Clin Endocrinol* 1979; **10:** 393–399.
12. Laron Z, Lazar L, Pertzelan A. Effect of human growth hormone treatment on head circumference of children with isolated GH deficiency. Submitted.
13. Niwa M, Sato Y, Uchiyama F, Ono H, Yamashita M, Kitaguchi T. Chemical synthesis, cloning and expression of genes for human somatomedin C (insulin like growth factor I) and ^{59}val-somatomedin C. *Ann NY Acad Sci* 1986; **469:** 31–52.
14. Larson Z, Anin S, Klipper-Aurbach Y, Klinger B. Effects of insulin-like growth factor on linear growth, head circumference, and body fat in patients with Laron-type dwarfism. *Lancet* 1992; **339:** 1258–1261.
15. Laron Z, Anin S, Klinger B. Long-term IGF-I treatment of children with Laron syndrome. In *Lessons from Laron Syndrome (1966–1992): A Model of GH and IGF-I Action and Interaction*. Pediatric and Adolescent Endocrinology, Vol. 24, edited by Laron Z, Parks JS, Karger, Basel, 1993, pp. 226–236.
16. Goodman HG, Grumbach MM, Kaplan SL. Growth and growth hormone. *N Engl J Med* 1968; **278:** 57–68.
17. Burt L, Kulin HE. Head circumference in children with short stature secondary to primary hypothyroidism. *Pediatrics* 1977; **59:** 628–630.
18. Kaplan SL. Growth: normal and abnormal. In *Pediatrics*, 18th edn, edited by Rudolph AM, Appleton & Langh, Norwalk. 1987; pp. 75–91.
19. Laron Z. Comparison between hGH and IGF-I. In *Lessons from Laron Syndrome (1966–1992): A Model of GH and IGF-I Action and Interaction*. Pediatric and Adolescent Endocrinology, Vol. 24, edited by Laron Z, Parks JS, Karger, Basel, 1993, pp. 319–321.
20. Daughaday WH. Are there direct, non-mediated effects of hGH? In *Lessons from Laron Syndrome (1966–1992): A Model of GH and IGF-I Action and Interaction*. Pediatric and Adolescent Endocrinology, Vol. 24, edited by Laron Z, Parks JS, Karger, Basel, 1993, pp. 338–345.
21. Murphy LJ, Bell GI, Friesen HG. Tissue distribution of insulin-like growth factor I and II messenger ribonucleic acid in the adult rat. *Endocrinology* 1987; **120:** 1279–1282.
22. Rotwein P, Burgess SK, Milbrandt JD, Krause JE. Differential expression of insulin-like growth factor genes in rat central nervous system. *Proc Natl Acad Sci USA* 1988; **85:** 265–269.
23. Hasselbacher GK, Schwab ME, Pasi A, Humbel RE. Insulin-like growth factor II (IGF-II) in human brain: regional distribution of IGF-II and of higher molecular mass forms. *Proc Natl Acad Sci USA* 1985; **82:** 2153–2157.
24. Sara VR, Carlsson-Skwirut C, Andersson C, Hall E, Sjogren B, Holingren A, Jornvall H. Characterization of somatomedins from human fetal brain: identification of a variant form of insulin-like growth factor I. *Proc Natl Acad Sci USA* 1986; **83:** 4904–4907.

25. Sara VR, Carlsson-Skwirut C, Bergman T, Jornvall H, Roberts PJ, Crawford M, Hakansson LN, Civalero I, Nordberg A. Identification of gly-pro-glu (GPE), the amino terminal tripeptide of insulin-like growth factor I which is truncated in brain, as a novel neuroactive peptide. *Biochem Biophys Res Commun* 1989; **165:** 766–771.
26. Lenoir D, Honegger P. Insulin-like growth factor I (IGF-I) stimulates DNA synthesis in fetal brain cell cultures. *Brain Res* 1983; **283:** 205–213.
27. McMorris FA, Smith TM, DeSalvo S, Furlanetto RW. IGF-I/somatomedin C: a potent inducer of oligodendrocyte development. *Proc Natl Acad Sci USA* 1986; **83:** 822–826.
28. Mill JF, Chao MV, Ishii DN. Insulin, insulin-like growth factor II and nerve growth factor effects on tubulin mRNA levels and neurite formation. *Proc Natl Acad Sci USA* 1985; **82:** 7126–7130.
29. Frankel JJ, Laron Z. Psychological aspects of pituitary insufficiency in children and adolescents with special reference to growth hormone. *Israel J Med Sci* 1968; **4:** 953–961.
30. Galatzer A, Aran O, Beit-Halachmi N, Nofar E, Rubitchek J, Pertzelan A, Laron Z. The impact of long term therapy by a multidisciplinary team on the education, occupation and marital status of growth hormone deficient patients after termination of therapy. *Clin Endocrinol* 1987; **27:** 191–196.
31. Galatzer A, Aran O, Nagelberg N, Rubitzek J, Laron Z. Cognitive and psychosocial functioning of young adults with Laron syndrome. In *Lessons from Laron Syndrome (1966–1992): A Model of GH and IGF-I Action and Interaction.* Pediatric and Adolescent Endocrinology, Vol. 24, edited by Laron Z, Parks JS, Karger, Basel, 1993, pp. 53–60.
32. Laron Z. Deficiencies of growth hormone and somatomedins in man. In *Special Topics in Endocrinology and Metabolism*, Vol. 5, edited by Cohen MP, Foa PP, Liss, New York, 1983; pp. 149–199.

Discussion of Presentation

ZVI LARON
(Edited by D.A. Snider and Salvador Castells)

R. McVie: Do you have a standard error of the mean of the *z* scores for the IQ changes for the GH-deficient children who were treated with GH?

Z. Laron: Yes, it was significant compared to the population.

P. Wilton: I think you have clearly shown that your population of LTD children had a rather low IQ, and you are well aware that the population found in Ecuador, on the contrary, has an above normal IQ. You showed data about the socio-economic conditions in your patients, but I really didn't understand it. Does a low socio-economic condition explain the difference in performance between your patients and the patients in Ecuador?

Z. Laron: I don't think that the interpretation of data provided by the Ecuadorian study is correct. These patients did not undergo repetitive IQ tests, nor psychological tests compared to a normal population of the same background. I have discussed this with Dr Rosenbloom. If they do the proper tests, then I think they will find that they have a deficit.

P. Wilton: So what about the socio-economic status in your families?

Z. Laron: There may be a certain influence, but it is not very strong because GH-deficient patients who had the same findings had better performance.

P. Wilton: You showed on two slides—two children with hGH deficiency—that convinced me that treated before four years of age increased in head circumference, but not if above eight years. You also showed that your LTD patients, even if they were 10 or 14 years of age, also increased in head circumference. How do you explain the difference?

Z. Laron: There were six children with GH deficiency due to gene deletion, and there was a possible difference, but the number is much too small. There was also some difference within patients. The cranial sutures need to be open to let the brain grow—it cannot grow if the sutures are closed. The LTD patients are the extreme of IGF-1 deficiency. The other factor is that the hGH-deficient children were treated at an earlier age. They started treatment between the ages of two and four years, so that we cannot really compare the data.

P. Wilton: But you showed us that if you treat GH-deficient patients after the age of eight, they don't increase in head circumference. You have two LTD patients 10 and 14 years old with open sutures—is that true?

Z. Laron: Correct. You see these patients were not hGH gene deletions, who were treated; they probably have acquired deficiency, and they have closed cranial sutures.

13 Long-term Effects of Recombinant Human Growth Hormone (rhGH) on Children with Down's Syndrome with Short Stature

SALVADOR CASTELLS,* ISLANDE BEAULIEU AND KRYSTYNA E. WISNIEWSKI

State University of New York Health Science Center at Brooklyn, Children's Medical Center of Brooklyn, Institute for Basic Research in Developmental Disabilities, New York, USA

Linear growth in children with Down's syndrome (DS) (trisomy 21) is often significantly retarded in comparison with that in normal children (1). Growth retardation usually begins several months after birth and extends through adolescence (2). Children with DS manifest microcephaly several months after birth (3), accompanied by retardation of brain growth and hypothalamic abnormalities (4). A chromosomal defect—trisomy of chromosome 21—appears to be the cause of DS, with the abnormalities in the central nervous system (CNS) of DS subjects thought to be the result of the regulation on dosage effects of one or more genes on chromosome 21. Some of these genes that may be involved in causing some of the CNS abnormalities and the accelerating aging found in persons with DS have been found to be located on chromosome 21 (5, 6). On the other hand, the severe growth retardation of children with DS has not been localized in any gene of chromosome 21. We have postulated that the growth retardation in children with DS may be caused by a dysfunction of the hypothalamic–pituitary axis (7), secondary to the diffuse neuropathological and biochemical abnormalities found in the brains of DS children (4).

In a recent study of 20 DS children younger than six years of age with severe growth retardation without congenital heart disease (CHD), we found poor hypothalamic responses to L-dopa and clonidine stimulation for serum growth hormone (GH) concentrations (7). In another study, we observed the effects of recombinant human growth hormone (rhGH) on children with DS

*Address for correspondence: State University of New York Health Science Center at Brooklyn, Department of Pediatrics, Box 49, 450 Clarkson Avenue, Brooklyn, NY 11203, USA.

Growth Hormone Treatment in Down's Syndrome. Edited by S. Castells and K.E. Wisniewski

for a period of one year. Therapy with rhGH resulted in a significant increase in annual growth rate and head circumference in children with DS (8).

In the present study, we treated 12 children with DS, somatic growth retardation, microcephaly, and partial GH deficiency without CHD with rhGH for more than one year. Growth velocity for the years of treatment was compared with that during the year before treatment.

METHODS

The 12 children studied were demonstrated by chromosomal analysis to have trisomy 21 and were referred to us from clinical and private sources. None of the DS children studied had CHD; children with CHD without trisomy of chromosome 21 also manifest growth retardation (9, 10). The following criteria were required for DS children to be included in the treatment protocol: (a) growth retardation with a growth velocity less than normal for age, for one year before treatment with rhGH; (b) low (< 10 ng/ml) peak serum GH concentration after stimulation with L-dopa or clonidine tests, or both; (c) bone age < −1.5 SD for chronological age; (d) no clinical evidence of CHD; (e) no evidence of psychological deprivation or abuse; and (f) appropriate nutrition for age. Each child studied was cared for by a stable family at home. None had a family history of short stature, endocrine disorders, mental retardation or consanguinity. All children were fed at home by their parents and given appropriate diets for age.

ANTHROPOMETRIC MEASURES

Height was measured in children older than two years of age, and length in those younger than two years of age. Height was measured with a wall-mounted stadiometer and recumbent length was measured with the child lying on his or her back on a measuring board, with one person holding the child's head against a board perpendicular to the measuring board. Head circumference (i.e. maximal occipitofrontal circumference) was measured using a paper tape. Growth velocity for height one year before treatment was obtained from growth charts provided by the patient's own pediatricians or by our own growth charts when subjects were followed by us before therapy. For all measurements, the average of three consecutive measurements was used as a final value.

The mean heights and weights for normal children were obtained from the National Center for Health Statistics growth curves for children (11). Head circumference for normal children were obtained from results previously published by Lowrey (12). Bone age for children two years of age and older was determined by obtaining roentgenograms of the left hand and wrist and comparing them with roentgenograms of the left hand and wrist of children of the same chronological age and sex (13). For children younger than two years of age, the bone age was measured by obtaining roentgenograms of the left hand and wrist, knee and ankle and comparing them with roentgenograms of normal children of the same chronological age and sex (14).

Figures 1–4 indicate the heights of our research subjects in comparison with the heights of normal children before and during treatment with rhGH. They were drawn with the computer program GrowTrak from the National Cooperative Growth study (Genentech, South San Francisco, CA, USA). Heights were obtained every three months. Figures 5 and 6 demonstrate the increases in head circumference during rhGH treatment in comparison with standard

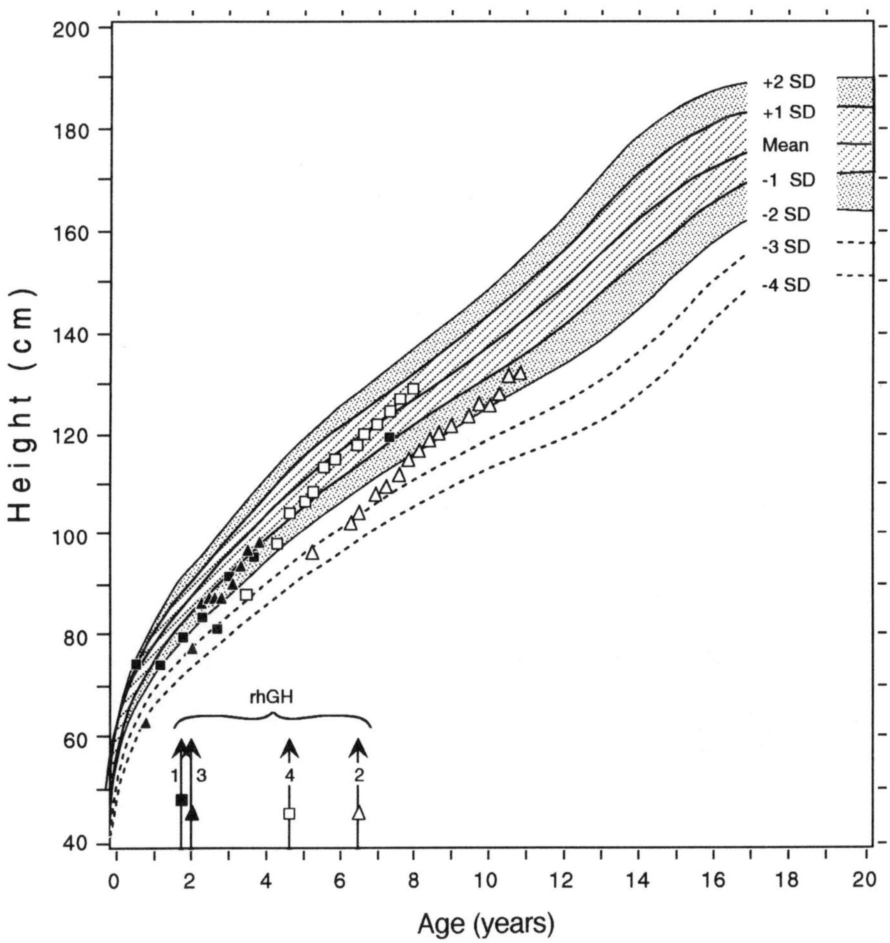

Figure 1. Height of four boys with DS receiving treatment with rhGH from three to five years

normal head circumference graphs for males and females. The head circumference figures were drawn by hand.

LABORATORY MEASUREMENTS

A complete blood cell count, urinalysis and biochemical analysis of serum sodium, potassium, chloride, carbon dioxide, glucose, urea nitrogen, creatinine, uric acid, calcium, phosphorus, alkaline phosphatase, total bilirubin, total protein, albumin, cholesterol, glutamic oxalacetic acid, transaminase, glutamic pyruvic transaminase and lactate dehydrogenase, by the SMA-12 and

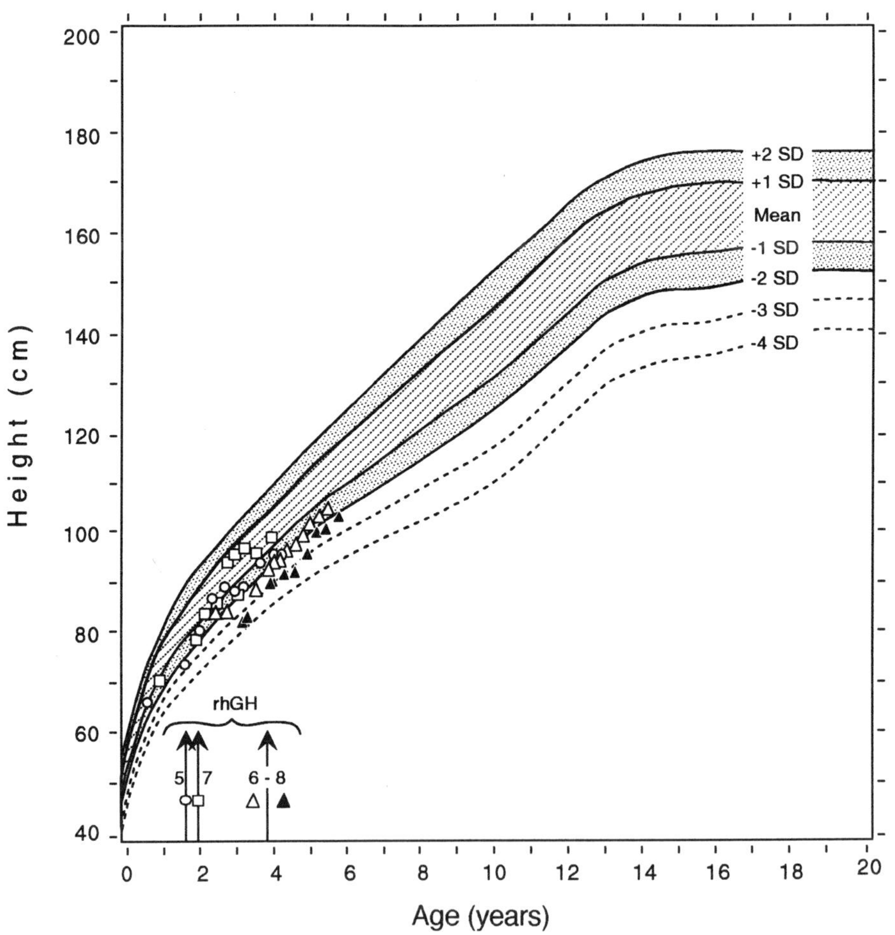

Figure 2. Height of four girls with DS receiving treatment with rhGH from two to four years

-6 by the Kodak Ektachem-700 analyzer (Eastman Kodak Co., Rochester, NY, USA) were performed at three-month intervals during the period of study. Hemoglobin A_{1c} was also measured at three-month intervals using the method of Bio-Rad Laboratories, Clinical Division, Hercules, CA for detection of hyperglycemia. No significant changes were detected during the period of treatment in any of the children with DS. Serum concentrations of thyroxine (T_4), 3,5,3′-triiodothyronine (T_3) and thyrotropin (TSH) were determined at three-month intervals during the period of study, because of the high incidence of thyroid disease in DS children. All the thyroid hormones were determined by standard radioimmunoassay (RIA) kits: TSH—Irma-Count, TSH with monoclonal antibodies (Diagnostic Product Corporation, Los Angeles, CA,

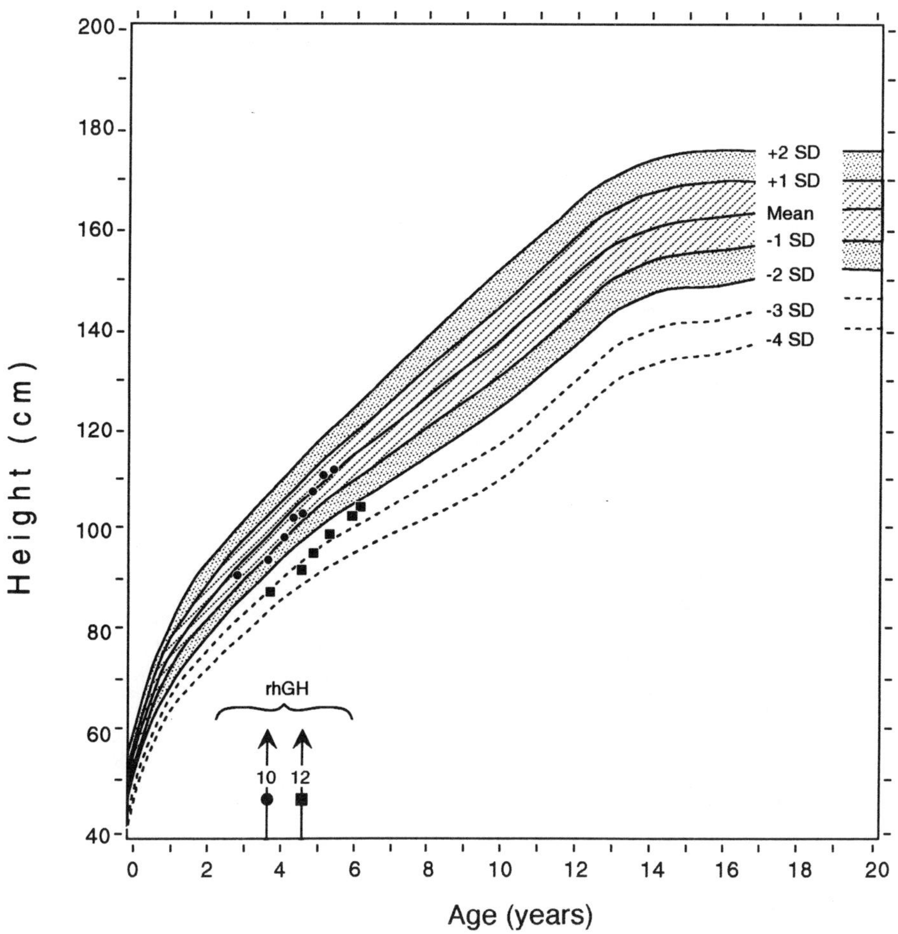

Figure 3. Height of two girls with DS receiving treatment with rhGH for one year

USA); T_3—Gamma Coat TM, [^{125}I]T_3 radioimmunoassay kit (Baxter Travenol Diagnostics, Inc., Cambridge, MA, USA); T_4—T_4 Gamma Coat TM, ^{125}I-labeled free total T_4 radioimmunoassay kit (Baxter Travenol Diagnostics, Inc.). Intra-assay and inter-assay coefficients of variation for TSH, T_3 and T_4 were 2.7% and 5.7%; 4.9% and 5.8%; 5.0% and 5.3%, respectively. GH and somatomedin-C or insulin-like growth factor 1 (IGF–1) were determined by immunoassay: GH—Quintitope hGH RIA kit (Kallestad, Inc., Austin, TX, USA); somatomedin-C (Nichols Institute Diagnostics, San Juan Capristrano, CA., USA) and IGF-1 were measured by Metpath RIA after acid extraction and purification on octadecasilyl silica cartridges (15). Intra-assay and inter-assay coefficients of variation for GH, somatomedin-C and IGF-1

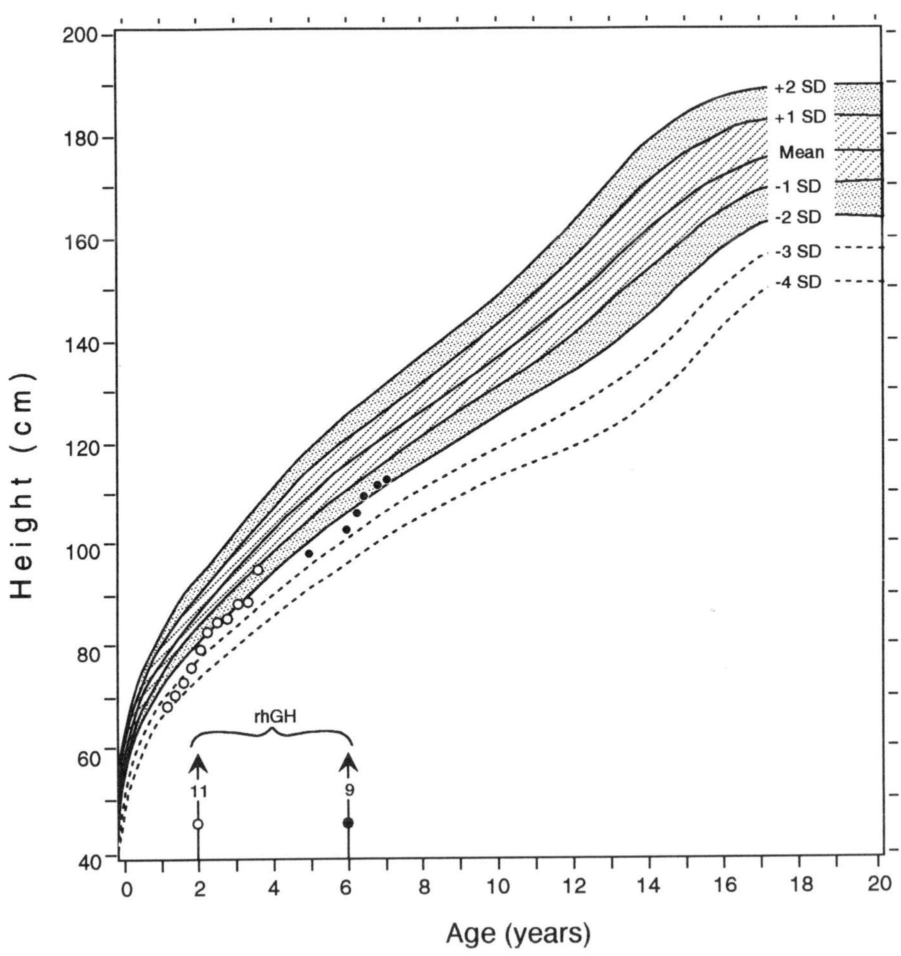

Figure 4. Height of two boys with DS receiving treatment with rhGH for one year

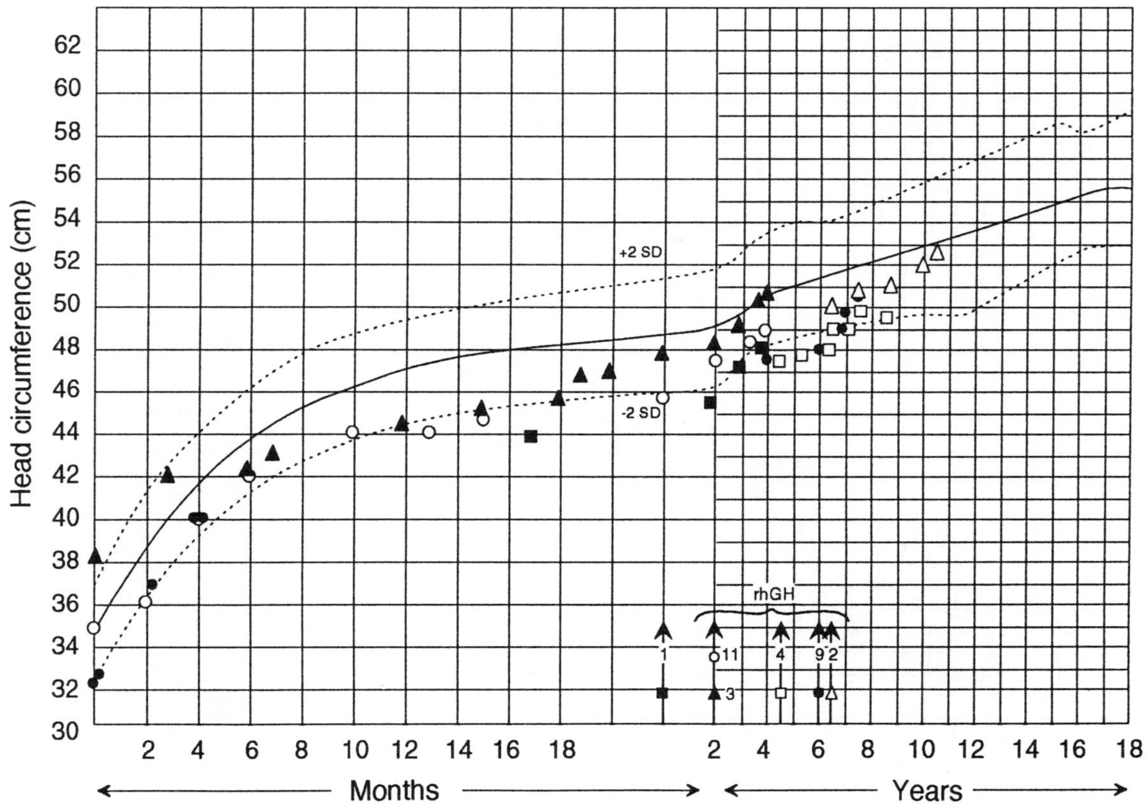

Figure 5. Head circumference of six boys with DS receiving treatment with rhGH

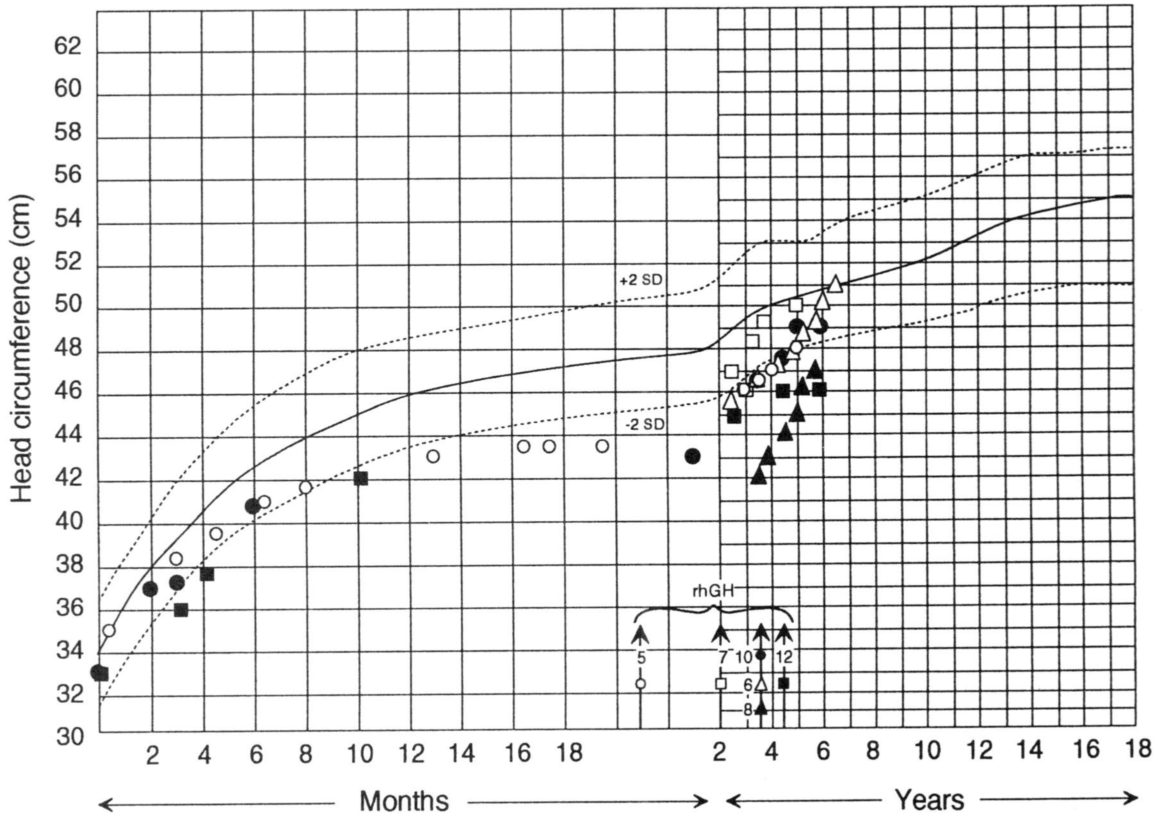

Figure 6. Head circumference of six girls with DS receiving treatment with rhGH

were 3.0% and 9.5%; 11.8% and 11.9%; and 10.0% and 12%, respectively. Each assay was performed in duplicate, and samples from each subject were assayed in the same batch to reduce inter-assay variability.

L-dopa and clonidine stimulation tests for the serum GH concentrations were performed after an overnight fast, with the patient at complete rest. L-dopa was administered orally according to body weight: below 15 kg, 125 mg tablet was given; between 15 and 30 kg, 250 mg tablet was given. Blood samples were drawn before and at 20, 40, 60, 90 and 120 min after L-dopa administration. Clonidine was administered orally, by a single dose of 0.15 mg/m^2 body surface area. Blood samples were drawn at 0, 30, 60, 90 and 120 min for determination of GH. An indwelling catheter was placed in a vein at initiation of each test. An anticoagulant (heparin) was used in small amounts to flush the catheter at the start and during the test, if needed. A peak response to L-dopa or clonidine of more than 10 ng/ml was considered normal. The peak and ▲-maximum increase (maximum increase = peak serum GH concentration − basal serum GH concentration) values of serum GH concentration after L-dopa and clonidine were compared with those of seven normal children (five males and two females, aged 2 years 7 months to 13 years) with short stature. The mean bone age SDS for the seven children used as controls was −0.18. In all control children, serum GH concentrations after L-dopa and clonidine administrations were >10 mg/ml.

Formal written consent was obtained before the tests were performed. The studies were approved by both Institutional Review Boards at the State University of New York Health Science Center at Brooklyn and the New York State Office of Mental Retardation and Developmental Disabilities, Institute for Basic Research in Developmental Disabilities on Staten Island.

TREATMENT

Methionyl rhGH, 0.1 mg/kg, was administered three times a week subcutaneously. One or both parents were trained to administer the drug in our clinic. Administration of the drug by one parent at home was started after one or two weeks of outpatient instruction and training. Ten children with DS receiving treatment with rhGH were seen at intervals of three months in our clinic; two other children were followed by outside institutions, and their anthropometric measurements were made available to us.

STATISTICAL ANALYSIS

For normally distributed data, statistical analyses were performed with Student's *t*-test for paired and unpaired analyses. The data from L-dopa and clonidine stimulation tests demonstrated a far from normal distribution. Two non-parametric statistical methods, more appropriate than the parametric methods, for use with data not conforming to the normal distribution, were used to analyze the results. The median and Kruskal-Wallis methods were used to analyze the results and compare the results among groups (control versus DS children).

RESULTS

ENDOCRINE RESULTS (Tables 1 and 2)

No initial evidence of hypothyroidism was found in any of the subjects studied, as documented by the normal concentrations of T_4, T_3 and TSH before the initiation of rhGH treatment. In 10 subjects, serum levels of prolactin, luteinizing hormone (LH), follicle-stimulating hormone (FSH) and

Table 1. Serum GH concentrations (ng/ml) during L-dopa stimulations test in children with DS and controls

Patient no.	L-Dopa		
	Basal	Peak	▲ Max.
Control			
1	0.4	11.8	11.4
2	1.0	12.9	11.9
3	0.8	30.2	29.4
4	1.8	14.7	12.9
5	21.8	18.7	3.1
6	3.9	25.7	41.8
7	12.9	14.4	1.5
Mean ± SD	6.1 ± 8.2	18.3 ± 7.0	12.3 ± 11.1
DS			
1	2.1	2.2	0.1
2	5.6	3.8	−1.8
3	8.6	8.3	−0.3
4	5.3	29.3	24.0
5	0.9	13.8	12.9
6	1.7	1.6	−0.1
7	1.6	7.3	5.7
8	8.5	13.9	5.4
9	2.9	6.2	3.3
10	5.8	11.2	5.4
11	4.2	7.3	3.1
12	1.5	1.7	0.2
Mean ± SD	4.1 ± 2.7	8.9 ± 7.7	4.8 ± 7.2
Kruskal–Wallis *p*	n.s.	$p < 0.007$	n.s.

estrogen or testosterone were at normal prepubertal concentrations before and during the period of treatment. (No follow-up values are available for subjects 1 and 3.) In the controls, the mean ± SD serum GH concentration during the L-dopa stimulation test during the basal period was 6.1 ± 8.2, peak response was 18.3 ± 7.0, and ▲ maximum response was 12.3 ± 11.1 ng/ml, and in the subjects with DS the basal serum GH concentration was 4.1 ± 2.7, peak response was 8.9 ± 7.7, and ▲ maximum was 4.8 ± 7.2 ng/ml (peak response between control versus DS, $p < 0.007$) (Table 1). The mean ± SD serum GH concentration during the clonidine stimulation test in controls during the basal period was 3.2 ± 2.8, peak response was 20.5 ± 6.6, and ▲ maximum response was 17.3 ± 5.6 ng/ml; in the subjects with DS, basal was 6.1 ± 7.7, peak response was 9.9 ± 6.4, and ▲ maximum was 3.8 ± 10.0 ng/ml (peak

Table 2. Serum GH concentrations (ng/ml) during clonidine stimulation test in children with DS and controls

Patient no.	Clonidine		
	Basal	Peak	▲ Max.
Control			
1	3.1	20.6	17.5
2	0.8	19.0	18.2
3	0.9	26.8	25.9
4	1.9	10.4	8.5
5	8.3	28.2	19.4
6	5.6	24.7	19.1
7	1.7	14.0	12.3
Mean ± AD	3.2 ± 2.8	20.5 ± 6.6	17.3 ± 5.6
DS			
1	1.4	6.0	4.6
2	15.4	7.9	−7.5
3	0.8	25.3	24.5
4	25.1	10.0	−16.1
5	10.3	11.9	1.6
6	2.3	6.6	4.3
7	1.8	3.6	1.8
8	1.8	4.0	2.2
9	3.3	18.7	15.4
10	5.4	9.7	4.3
11	3.8	10.2	2.2
12	0.9	4.5	3.6
Mean ± SD	6.1 ± 7.7	9.9 ± 6.4	3.8 ± 10.0
Kruskal–Wallis p	n.s.	$p < 0.04$	$p < 0.04$

and ▲ maximum responses between control versus DS was $p < 0.04$ and $p < 0.04$, respectively) (Table 2). Five children with DS (nos 1, 2, 6, 7, 12) had serum concentrations of GH peaks with both L-dopa and clonidine stimulation tests below 10 ng/ml. Eight subjects had a peak serum concentration after L-dopa below 10 ng/ml, and seven subjects had peak serum concentration after clonidine below 10 ng/ml.

ANTHROPOMETRIC DATA AND RESPONSE TO THERAPY (Tables 3–6)

The 12 children with DS—six males and six females—were short for age and microcephalic before therapy was started. Therapy was started between ages 1 year and 10 months and 6 years. Growth velocity per year before treatment was found to be less than the normal mean height velocity for age and sex (16). Subjects were placed in groups depending on the number of years of therapy with rhGH (Table 3). Three subjects who had been receiving rhGH for five

Table 3. Anthropometric measurements for the children studied

Length of hGH treatment (years)	Age (years; months)	Sex	Growth velocity (cm/year)						Normal mean height velocity for age[c]
			Before[b]	1st	2nd	3rd	4th	5th	
5 years									
1[a]	1;10	M	6.0	9.0	8.0	11.8	7.0	5.0	12;0
2	6;4	M	4.0	9.0	8.8	4.2	6.4	2.5	5;0
3	2;0	M	8.0	13.0	7.9	6.0	11.0	7.4	10;0
3 years									
4	4;6	M	4.0	14.5	6.5	7.0			7;8
5	1;9	F	4.2	9.7	10.7	8.0			11;9
6	3;10	F	4.0	9.1	7.0	8.0			7;8
2 years									
7	2;10	F	4.5	13.6	6.8				10;0
8	3;10	F	5.0	10.3	7.2				7;8
1 year									
9	6;0	M	5.1	9.7					5;0
10	2;0	M	9.6	14.0					10;0
11	3;10	F	4.0	9.0					7;8
12	4;5	F	5.0	9.0					7;8

[a]Height measurements obtained from outside institutions.
[b]Height data obtained from growth charts provided by private pediatricians or by our own measures of height when patients were followed by us before the initiation of therapy.
[c]Obtained from Lifshitz F, Zandsberg S. Disorders of growth. In *Pediatric Endocrinology* edited by Lifshitz F, Dekker, New York, 1985; pp. 11–12.

Table 4. Bone age in children with DS before and during treatment with hGH; chronological age (CA) and Bone age (BA) (years;months)

Length of hGH treatment (years)	Sex	Before		1st		2nd		3rd		4th		5th	
		CA	BA	CA	BA	CA	BA	CA	BA	CA	BA	CA	BA
5 years													
1[a]	M	1;5	0;9	2;10	2;0					7;6	6;9	8;0	7;10
2	M	6;0	3;6	7;2	5;0	8;4	6;0	9;7	8;0				
3	M	1;7	1;0									6;0	5;5
3 years													
4	M	4;4	3;0	5;7	4;6	6;8	6;0						
5	F	1;6	0;11	2;3	2;0	3;3	3;2						
6	F	3;7	2;0	4;2	3;6	5;9	5;9						
2 years													
7	F	1;4	1;3	1;8	1;8	2;7	2;0						
8	F	3;7	1;6	4;7	2;6	6;0	4;2						
1 year													
9	M	4;9	4;0	6;0	5;6								
10	F	3;8	2;0	4;11	3;6								
11	M	1;10	0;9	3;0	2;0								
12	F	4;5	3;0	6;1	6;0								

Table 5. Head circumference before and during treatment with hGH for children studied

Length of treatment (years)	Head circumference (cm)						
	Pre	1st	2nd	3rd	4th	5th	6th
5 years							
1[a]	44.5	46.5	47.0				
2	48.5	51.0	51.5	51.5	52.5	52.5	
3[a]	46.5	50.5	51.0	51.0	51.0	51.0	
3 years							
4	47.0	48.0	49.0	49.5			
5	43.5	46.5	46.8	47.5			
6	46.8	49.5	49.5	49.5			
2 years							
7	44.5	47.5	49.0				
8	42.0	45.0	47.0				
1 year							
9	48.5	49.5					
10	46.4	48.0					
11	46.5	48.5					
12	47.0	47.0					

[a]Head circumference measurements obtained from outside institutions.

Table 6. Head circumference growth rate per year before and during treatment with hGH for children studied

Length of treatment (years)	Growth rate per year (cm)						
	Pre	1st	2nd	3rd	4th	5th	6th
5 years							
1[a]		2.0	0.5	0.5			
2		2.5	0.5	0.0	1.0	0.0	
3[a]		4.0	3.0	1.0	0.5	0.0	
3 years							
4		1.0	1.0	1.0	0.5		
5		0.0	3.0	0.3	0.7		
6		1.3	2.7	0.0	0.0		
2 years							
7		0.0	3.0	1.5			
8		1.0	2.0	2.0			
1 year							
9		0.5	1.0				
10		1.4	1.6				
11		0.5	2.0				
12		1.0	1.0				

[a]Head circumference measurements obtained from outside institutions.

Table 7. Somatomedin-C[a] concentrations before and during hGH treatment in children studied

Length of treatment (years)	IGF-1[a] (units/ml)						
	Pre	1st	2nd	3rd	4th	5th	6th
5 years							
1							
2	0.83	1.05	1.16	2.75	3.30		
3	0.42	3.5	1.4				
3 years							
4	0.68	2.0	3.10				
5	0.50	2.29	1.40				
6	0.50	0.83	2.62	0.95			
2 years							
7	0.56	2.56					
8	0.67	1.71	1.34				
1 year							
9	0.76						
10	1.04						
11	0.65	1.77					
12	0.37	0.86					

[a]Concentrations of IGF-1 or somatomedin-C were determined by RIA of unextracted serum by Metpath RIA.

Table 8. IGF-1 plasma concentrations during hGH treatment in children studied

Length of treatment (years)	IGF-1[a] (units/ml)					
	Pre	1st	2nd	3rd	4th	5th
5 years						
1						468.6
2						
3				171.4		
3 years						
4				287.8		
5				303.2		
6					124.0	
2 years						
7						
8						
1 year						
9			174.0		258.5	
10			391.5			
11						
12						

[a]Concentrations of IGF-1 were determined after acid extraction of serum and purification on octadecasilyl silica cartridges by Metpath RIA.

or more years experienced a significant increase in growth velocity during the first year of treatment over the growth rate before therapy. Growth has continued for the five years of treatment. In three additional subjects who had been receiving rhGH for three years, the growth rate per year was markedly decreased during the year before therapy and increased to normal or above normal each year during the three years of therapy. Two subjects who had been receiving rhGH for two years had decreased growth velocity during the year before therapy and significant increases in growth velocity each year of therapy. Four subjects who had been receiving rhGH for only one year experienced decreased growth velocity during the year before therapy and significant increases in growth velocity during each year of therapy. The growth velocity for children with DS receiving rhGH in comparison with that for normal subjects of the same sex is given in Figs 1–4. Bone age, measured every 12 months, did not advance during the period of treatment with rhGH in relationship to chronological age (Table 4).

Head circumferences were in the range of microcephaly in all the DS children before treatment with rhGH. Increases in head circumference and reversibility to normocephaly were noticed in all the children with DS after therapy with rhGH (Tables 5 and 6 and Figs 5 and 6).

LABORATORY TESTS

Complete blood count, urinalysis and routine serum chemistry values were normal before and during rhGH treatment. Plasma hemoglobin A_{1c} concentration remained normal during treatment with rhGH in the 10 children tested.

In all cases, plasma concentrations of IGF-l were in the low range of normal for age and gender before initiation of therapy, and increased significantly during treatment (Tables 7 and 8).

DISCUSSION

Growth retardation and microcephaly in individuals with DS have been recognized, but their causes have been unknown. We have postulated that hypothalamic dysfunction, primary or secondary to pathological and/or biochemical abnormalities in neurotransmitters in the brain and hypothalamus, might play a role in the growth retardation and microcephaly of children with DS (7, 17). Recently, morphogenic abnormalities in the hypothalamus of DS subjects, especially diminished numbers of neurons in the arcuate and ventromedial nuclei, have been reported (18). Also, biochemical abnormalities in the hypothalamus of DS children have been reported. Measurements of serum GH concentrations during L-dopa, clonidine and 24 h integrated growth hormone concentrations (ICGH) indicate a biochemical hypothalamic dysfunction in DS as the main cause for neurosecretory deficiency of GH (7, 8).

For the present study, we used two drugs that stimulate hypothalamic release of GHRH: L-dopa and clonidine (19–24). Mean peak serum concentrations of GH after L-dopa and clonidine stimulation tests were significantly lower in individuals with DS than in controls, suggesting a hypothalamic dysfunction to L-dopa and clonidine stimulations. Testing with L-dopa for serum GH concentrations revealed that eight children had a peak response less than 10 ng/ml, and testing with clonidine indicated that seven children had a peak response less than 10 ng/ml (Tables 1 and 2). Five children (nos 1, 2, 6, 7, 12) had serum concentrations of GH peaks with L-dopa and clonidine stimulation tests < 10 ng/ml, indicating a deficiency in GH. The present authors regard 10 ng/ml as the arbitrary cut-off for deficiency of GH. The current opinion of most clinicians is that the discriminator of 10 ng/ml for peak serum GH responses to provocative stimuli is arbitrary (25), but most would agree with the diagnosis of GH deficiency in these five DS children. Six children (nos 3, 4, 8, 9, 10, 11) had a discrepancy in the response of peak serum concentrations of GH to L-dopa and clonidine. Diagnosis of GH deficiency in these children is more difficult. Evidence suggesting a state of diminished GH secretion in limited or disparate responses to different stimuli has been reported (26). A single dose of clonidine (0.15 mg/m^2) produces a greater peak GH response than does induced hypoglycemia or arginine and might give rise to false positive responses (27). The discrepancy in peak increases in serum GH concentrations after L-dopa and clonidine stimulation tests in the six DS children is suggestive of the difference in the intensity and the biochemical abnormality of neurotransmitters in the CNS that stimulate the hypothalamic–pituitary axis. Thus the hypothalamic dysfunction that is conducive to GH deficiency and short stature in persons with DS appears to be due to a mechanism indirectly related to the gene dosage and/or abnormalities in trisomy 21.

We have recently reported a normal serum concentration of GH response to GHRH in children with DS, indicating a normal pituitary response (normal somatotrope response to GHRH) (28) and further suggesting hypothalamic dysfunction.

Administration of rhGH to children with DS produced significant increase in growth velocity during the first and subsequent years of treatment, in comparison with the year before treatment. An increase in head circumference was also observed. Microcephaly has been associated with defective GH secretion (29, 30) and is a well-known finding in DS (3). Early correction of GH deficiency in children with DS served to promote head circumference growth, as it does in children with microcephaly associated with GH deficiency (30).

Plasma concentrations of IGF-1 have been found to be decreased in children with GH deficiency (31), but there is an overlap in plasma concentrations of IGF-1 in normal children and in children with GH deficiency (32). Plasma concentrations of IGF-1 are also influenced by chronological age (32), degree of sexual maturation (33, 34), and nutritional status (35). Plasma concentrations of IGF-1 before rhGH treatment were in the low range of normal for age and gender, as would be expected in individuals with GH deficiency. Our study subjects with DS were all of prepubertal age and had good nutrition, but because they were very young (less than six years of age) their pretreatment plasma concentrations of IGF-1 were difficult to compare with these values in controls.

Therapy with rhGH increased plasma concentrations of IGF-1 to higher levels than before therapy. Such increased plasma concentrations of IGF-1 in children with DS were maintained as long as therapy with rhGH was given. In previous work, we have found that in non-compliant individuals receiving rhGH treatment plasma concentrations reverted to the pretreatment values during the period of non-compliance (8).

In summary, mean peak serum concentrations of GH after L-dopa and clonidine stimulation tests were significantly lower in the subjects with DS than in controls. Five children with DS had peak serum concentrations of GH after stimulation with L-dopa and clonidine of below 10 ng/ml, which is considered diagnostic of GH deficiency. Six children with DS had a discrepancy in the peak serum concentrations of GH after each stimulus. We interpreted this discrepancy to be suggestive of neurosecretory dysfunction resulting from biochemical hypothalamic abnormalities.

All 12 children with DS, with diminished growth velocity before treatment with rhGH, responded with a marked increase in growth velocity during treatment with rhGH. Significant growth continued during the five years the patients were treated. Head circumferences have also increased in all of these individuals. We conclude that rhGH therapy can result in a significant increase in growth rate for up to five years and an increase in head circumference.

No changes in complete blood count, urinalysis, or in serum concentrations of sodium, potassium, chloride, carbon dioxide, glucose, urea nitrogen, creatinine, uric acid, calcium, phosphorus, alkaline phosphatase, total bilirubin, total protein, albumin, cholesterol, glutamic oxaloacetic transaminase, glutamic pyruvic transaminase, lactate dehydrogenase, prolactin, LH, FSH, estradiol or testosterone, T_4, T_3, TSH or glycohemoglobin were found, suggesting that rhGH in children with DS does not produce any side effects that would be evident in these biochemical tests for up to five years.

Final height and head size can be improved with rhGH treatment that is started at an early age and maintained for a long period of time.

REFERENCES

1. Cronk C, Crocker AC, Pueschel SM *et al*. Growth charts of children with Down's syndrome: 1 month to 18 years of age. *Pediatrics* 1988; **81:** 102–110.
2. Pueschel SM. Growth, thyroid function, and sexual maturation in Down's syndrome. *Growth Genet Horm* 1990; **6:** 1–5.
3. Wisniewski KE. Down's syndrome (DS) children, postnatal retardation of head and brain growth (birth–five years). In *Disorders of the Developing Nervous System: Changing Views on their Origins, Diagnoses, and Treatment*. Proceedings of the 1987 Albany Birth Defects Symposium XVIII, edited by Swann JW, Messer A, Liss, New York, 1987, p. 282.
4. Wisniewski KE. Down's syndrome children often have brain growth with maturation delay, retardation of growth and cortical dysgenesis. *Am J Med Genet* 1990; **7:** 274–281.
5. Epstein J. The consequences of altered gene dosage in trisomy 21. In *New Perspectives on Down's syndrome*, edited by Pueschel SM, Tingey C, Rynders JE, Crocker AC, Crutcher DM, Brookes, Baltimore, 1987; pp. 69–80.
6. Tanzi RE, Gusella JF, Watkins PC, Bruns GAP, St George-Hyslop P, Van Kauren HL, Patterson D, Pagan S, Kurnit DM, Nere RL. Amyloid B protein gene: cDNA, mRNA distribution, and genetic linkage near the Alzheimer locus. *Science* 1987; **235:** 880–884.
7. Castells S, Torrado C, Bastian W, Wisniewski KE. Growth hormone deficiency in Down's children. *J Intell Dis Res* 1992; **36:** 29–43.
8. Torrado C, Bastian W, Wisniewski KE, Castells S. Treatment of children with Down's syndrome and growth retardation with recombinant human growth hormone. *J Pediatr* 1991; **119:** 478–483.
9. Feldt RH, Strickler GB, Weidman WH. Growth of children with congenital heart disease. *Am J Dis Child* 1969; **117:** 573–579.
10. Mehziri A, Drash A. Growth disturbance in congenital heart disease. *J Pediatr* 1962; **61:** 418–429.

11. National Center for Health Statistics. *NCHS Growth Curves for Children, Birth to 18 years (US)*, National Center for Health Statistics, Hyattsville, MD, 1977.
12. Lowrey GH. *Growth and Development of Children*, 7th edn, Year Book Medical Publishers, Chicago, 1978; pp. 76–80.
13. Greulich WW, Pyle ST. *Radiographic Atlas of Skeletal Development of the Hand and Wrist*, 2nd edn, Stanford University Press, Stanford, CA, 1959.
14. Hernàndez M, Sànchez E, Sobradillo B, Rincòn JM, Narvaiza JL. A new method for assessment of skeletal maturity in the first 2 years of life. *Pediatr Radiol* 1988; **18:** 484–489.
15. Pintor C, Loche S, Cella SG, Baumann G. Correction and withdrawal of conclusion: a child with phenotypic Laron dwarfism and normal somatomedin levels. *N Engl J Med* 1990; **323:** 1485.
16. Lifshitz F, Zandesberg S. Disorders of growth. In: *Pediatric Endocrinology*, edited by F. Lifshitz, Dekker, New York, 1991; pp. 11–12.
17. Castells S, Torrado C, Bastian W, Wisniewski KE. Growth hormone deficiency in Down's syndrome. *Pediatr Res* 1988; **23:** 274A (abstract).
18. Wisniewski KE, Bobinski M. Hypothalamic abnormalities in Down's syndrome. In *The Morphogenesis of Down's Syndrome*, edited by Epstein CJ, Wiley–Liss, New York, 1991; pp. 153–167.
19. Abrams RL, Parker ML, Blanco S, Reichlein S, Daughaday WH. Hypothalamic regulation of growth hormone secretion. *Endocrinology* 1966; **78:** 605–613.
20. Boyd AE, Lebovitz HE, Pfeiffer JB. Stimulation of human growth hormone secretion by L-dopa. *N Engl J Med* 1970; **283:** 1425–1429.
21. Chihara K, Kashio Y, Kita T, Okimura Y, Kaji H, Abe H, Fujita T. L-Dopa stimulates release of hypothalamic growth-hormone releasing hormone in humans. *J Clin Endocrinol Metab* 1986; **62:** 466–473.
22. Donnadieu M, Evain-Brion D, Tonon MC, Vaudry H, Job JC. Variations of plasma growth hormone (GH)-releasing factor levels during GH stimulation tests in children. *J Clin Endocrinol Metab* 1985; **60:** 1132.
23. Delitala G, Maioli M, Pacifico A, Brianda S, Patermo M, Mannehi M. Cholinergic receptor control mechanisms for L-dopa, apomorphine, and clonidine-induced growth hormone secretion in man. *J Clin Endocrinol Metab* 1983; **57:** 1145–1149.
24. Gil-Ad I, Topper E, Laron Z. Oral clonidine as a growth hormone stimulation test. *Lancet* 1979; **ii:** 278–279.
25. Underwood LE, Sherman BM. Controversies in the treatment of short stature. In *Human Growth Hormone Progress and Challenges*, edited by Underwood LE, Dekker, New York, 1988; pp. 145–191.
26. Youlton R, Kaplan SL, Grumbach MM. Growth and growth hormone. IV. Limitations of the growth hormone response to insulin and arginine and the immunoreactive insulin response to arginine in the assessment of growth hormone deficiency in children. *Pediatrics* 1969; **43**: 989–1004.
27. Shulman D, Zamanillo J, Lowitt S. Effect of clonidine upon anterior pituitary function and plasma catecholamine concentrations in short children and adolescents. *J Pediatr Endocrinol* 1985; **4:** 211–216.
28. Castells S, Beaulieu I, Torrado C, Wisniewski KE, Zarny S, Gelato MC. Hypothalamic versus pituitary dysfunction in Down's syndrome (DS) as cause for growth retardation. *International Conference in Growth Hormone Treatment in Down's Syndrome*, New York, 10–11 October 1992 (abstract).
29. Dacou-Voutetakis C, Karpathios TH, Logothetis N, Constantindis M, Matsoniotis N. Defective growth hormone secretion in primary microcephaly. *Pediatrics* 1974; **85:** 498–502.
30. Spadoni GL, Cianfarani S, Bernardini S, Fabrizio V, Galasso C, Boscherini B.

Growth hormone treatment in children with sporadic microcephaly. *J Dis Child* 1989; **143:** 1282–1283.
31. Zapf J, Walter W, Froesch ER. Radioimmunological determination of insulin-like growth factors I and II in normal subjects and in patients with growth hormone disorders and extrapancreatic tumor hypoglycemia. *J Clin Invest* 1981; **68:** 1321–1330.
32. Bala RM, Lopatka J, Leung A, McCoy E, MacArthur RG. Serum immunoreactive somatomedin levels in normal adults, pregnant women at term, children at various ages, and children with constitutionally delayed growth. *J Clin Endocrinol Metab* 1981; **52:** 508–512.
33. Lund AM, Wilson DM, Wibbelsman CJ *et al.* Somatomedins in adolescence: a cross-sectional study of the effect of puberty on plasma insulin-like growth factor I and II levels. *J Clin Endocrinol Metab* 1983; **57:** 268–271.
34. Rosenfield RL, Furlanetto R, Bock D. Relationship of somatomedin-C concentration to pubertal changes. *J Pediatr* 1983; **103:** 723–728.
35. Clemmons DR, Klibanski A, Underwood LE. Reduction of plasma immunoreactive somatomedin-C during fasting in humans. *J Clin Endocrinol Metab* 1981; **53:** 1247–1250.

Discussion of Presentation

SALVADOR CASTELLS
(Edited by D.A. Snider)

A. Habid: As a clinician sometimes we see in children with Down's syndrome that the head size starts to go down before the length. Can we consider treating these kids with GH, before the length starts to go down just based on the head size?

S. Castells: This question has been brought to our attention several times by parents. Our protocol at this time calls only for treatment of short stature and does not recognize the problem of microcephaly. We may need to change our protocol in the future. We would have to convince our Committees on Human Investigations at our two institutions that there is enough evidence at the present time to recommend such a change.

N. Stahnke: Do you have any data on bone age development, during treatment, for, of course, all of these children given hGH treatment will grow faster. The point is do they really become taller?

S. Castells: The bone age does not accelerate with hGH treatment, and it follows chronological age.

14 Growth Retardation in Children with Down's Syndrome in Relation to Insulin-like Growth Factors and Growth Hormone Therapy

G. ANNERÉN,[1,2*] C. CARLSSON-SKWIRUT,[3]
J. GUSTAFSSON,[2] T. TUVEMO[2] AND V. R. SARA[3]
Departments of [1]Clinical Genetics and [2]Pediatrics, University Hospital, Uppsala, and [3]Karolinska Institute's Department of Pathology, Karolinska Hospital, Stockholm, Sweden

ABSTRACT

Between six months and three years of age growth velocity is markedly reduced in children with Down's syndrome (DS) as compared to healthy children. However, after three years of age it is almost normal. Thus growth retardation becomes pronounced during the period when growth hormone (GH) starts to regulate growth. There is no obvious GH deficiency in DS.

We have studied the insulin-like growth factor (IGF) system in DS. There is a selective IGF-1 deficiency from the age of two years in individuals with DS, which might account for the growth retardation. Seemingly, there is a delayed maturation with incomplete switching from production of the fetal form of IGF to production of the GH-regulated IGF-1 in DS. This might be due to a hypothalamic defect or a defect in the maturation of the GH–IGF system itself.

We have previously reported that children with DS are responders to GH, which gives increased growth velocity and normalized serum levels of IGF-1. We now report the long-term effects of hGH therapy from an ongoing study of 16 young children with DS, to be treated for three years from the age of six to nine months (mean age 7.4 months) with 0.1 U/kg body weight per day. The results after 30 months are presented. During hGH treatment the children with DS gained height during the first year, and then followed the growth rate of healthy Swedish children. As compared to growth charts for children with

*Address for correspondence: Department of Clinical Genetics, University Hospital, S-751 85 Uppsala, Sweden.

Growth Hormone Treatment in Down's Syndrome. Edited by S. Castells and K.E. Wisniewski

DS (*Pediatrics* 1988; **81**: 102) the mean height of the children started at the 50th centile, reaching the 95th centile after 24 months of treatment. Head circumference increased only slightly during the therapy, but not to the same extent as height.

GROWTH RETARDATION AND ENDOGENOUS GROWTH HORMONE PRODUCTION IN CHILDREN WITH DOWN'S SYNDROME

Down's syndrome (DS), which is the most common single specific cause of mental retardation (MR), is also characterized by growth retardation. The pattern of linear growth in children with DS is characteristic. The growth velocity is most markedly reduced between six months and three years of age, but is almost normal after three years of age (1, 2). Growth hormone (GH) regulates growth from the age of six to nine months postnatally by stimulation of production of insulin-like growth factor I (IGF-1). Thus the growth retardation becomes pronounced during the period when GH starts to stimulate growth. There is no obvious deficiency of GH in serum (3–5). Although we have found normal peaks of GH in short children with DS aged four to six years, the occurrence of suboptimal endogenous GH production due to a hypothalamic dysfunction has been suggested in DS (6) and a partial GH deficiency cannot be excluded.

GROWTH AND GROWTH REGULATION OF THE CENTRAL NERVOUS SYSTEM

The critical period for the growth and differentiation of the nervous system occurs during fetal and early postnatal life. During this period of time there is a rapid proliferation of the stem cells, whose daughter cells then migrate and differentiate into neurons and glia, forming their complex network of synaptic connections. During this development the selective programmed death of specific cells also occurs. Disturbance in this growth and differentiation process can result in permanent and irreversible defects in central nervous system (CNS) structures and influence the consequent function of the brain. While some plasticity is maintained within the nervous system throughout life, the developing glial cells and especially the neurons, which cease to proliferate after their differentiation, are particularly vulnerable during this early period of rapid growth. Understanding the mechanisms which regulate the proliferation and differentiation of the neurons and glia is thus of foremost importance in ensuring the optimal development of the CNS and its later function. In recent years it has become increasingly evident that brain growth and development are regulated by the complex interaction of several growth factors, such as fibroblast growth factor (FGF), nerve growth factor (NGF) and the insulin-like growth factors (IGFs). The present discussion will focus on the role of the IGFs, their binding proteins (IGFBPs) and receptors (IGFR), and raise the possibility of the involvement of these growth factors in the pathogenesis of DS.

BRAIN GROWTH IN RELATION TO THE INSULIN-LIKE GROWTH FACTOR SYSTEM

Both IGF-1 and IGF-2 are synthesized within the CNS, where expression of their genes is regulated in a developmentally and cell-specific manner (7–11). Using *in situ* hybridization, it has been possible to localize the cellular sites of their production. The IGF-1 gene is expressed in stem cells during periods of rapid proliferation, such as has been observed in the ependymal layer of the embryonic rat cerebral cortex, as well as in selective groups of neurons at specific stages of development especially during periods relating to myelination and the formation of synaptic connections. In contrast, IGF-2 gene expression is not detected in neuronal or glial cells but rather in the choroid plexus, leptomeninges and endothelial cells lining the cerebral vessels. IGFBP-2 is expressed in cells adjacent to those synthesizing IGF-2 in the choroid plexus. This localization supports the concept that the IGF-2–IGFBP-2 complex is formed extracellularly and that IGF-2 is transported in cerebrospinal fluid (CSF) as an IGFBP-associated complex to be distributed to target cells throughout the CNS. The IGFBPs are synthesized in the developing nervous system as well as in the adult brain, where the major forms in CSF appear to be IGFBP-2 and IGFBP-6, which both preferentially bind IGF-2 (10–14).

Both IGF-1 and IGF-2 are ligands activating the IGF-1R, which is the primary mediator of their growth-promoting actions (15). The IGF-1R is expressed early in development (16) and is widely distributed throughout cells of the CNS (9, 17, 18). Additionally, the expression of the IGF-1R increases during periods of increased IGF-1 expression, such as during rapid proliferation and synaptogenesis. The expression of the IGF-2R, which recognizes IGF-2 and not IGF-1, occurs early during embryogenesis and is also widespread in the CNS (19). However, apart from involvement in intracellular targeting of lysosomal enzymes, the function of this receptor is unclear (20). The expression of these genes is developmentally regulated and falls with maturation. In the adult brain, IGF-2 and IGFBP-2 continue to be synthesized in the choroid plexus, whereas IGF-1 displays very low or undetectable expression with the exception of restricted cell populations, such as the glomerular and mitral cells of the olfactory lobe (10). Of particular interest are the recent findings of Komoly *et al.* (21) and Lee *et al.* (22) that local expression can be switched back on again in response to injury in the CNS.

Both the IGF genes and their protein products have been characterized in the CNS. In the human fetal brain, both Ea IGF-1 and Eb IGF-1 mRNAs encoding for the two IGF-1 precursor proteins have been identified (23). The proteins produced from expression of the IGF-1 gene in the human brain have been isolated and characterized (24). The IGF-1 precursor protein is post-translationally modified by protease cleavage to result in two products: truncated IGF-1 and the tripeptide GPE (gly-pro-glu). These proteins fulfill separate CNS functions mediated by separate receptor mechanisms. The truncated IGF-1 is a potent neurotrophic factor (25) whereas the tripeptide GPE acts via an *N*-methyl-D-aspartate (NMDA) receptor and modulates neurotransmitter release (26). This amino-terminal truncation is functionally significant since the IGFBP binding site occurs over this region of the molecule. Therefore the direct biological activity of the truncated IGF-1 is enhanced since it is not bound by IGFBP and is readily available to the IGF-1R (25). In the human fetal brain, transcription from the IGF-2 gene appears to be initiated from the third promoter site in the gene (27). The protein products from expression of this gene in the human brain are either the mature form of IGF-2 or the higher molecular weight form, 'big' IGF-2, which represents proIGF-2 or a partially processed form of this precursor (24). Both forms of IGF-2 are present in the CSF, where they occur associated with IGFBP-2 and IGFBP-6 (13).

The production of the IGFs, IGFBPs and IGFRs is clearly developmentally regulated, and appears to occur in accordance with specific cellular growth stages (17, 28, 29). It is now well established that the expression of these genes is regulated by nutrition as well as by several hormonal factors. Nutrition appears to be one of the major regulators of IGF production during early development and it has been suggested that this may provide a major mechanism for nutritional influences on growth and development (30, 31). Several hormones, particularly GH, steroids and insulin, have also been demonstrated to regulate the production of the IGFs. For example, glucocorticoids are established as inhibitors of neuronal and glial cell formation and retard brain development (32). These effects appear to be mediated by glucocorticoid-induced inhibition of IGF-1 production at the level of gene transcription in both neuronal and glial cells. Of particular importance is the increasing recognition of the interregulation between several growth factors (33, 34). For example, Drago *et al.* (33) have demonstrated that FGF-induced proliferation of precursor cells isolated from the neuroepithelium of embryonic mice is dependent upon the production of IGF-1 by these cells.

The biological actions of the IGFs on the growth and differentiation of neural tissue have been well established both *in vitro* and *in vivo* (25, 35–41). IGF-1 and IGF-2 stimulate the proliferation of neuronal and glial cell precursors and this action is mediated via the IGF-1R. The IGFs are also involved in the regulation of neuronal and glial cell differentiation as well as the survival of these cells. Påhlman *et al.* (39) have recently demonstrated, for example, that during neuronal maturation IGF-1 shifts from promoting cell proliferation to potentiating neuronal differentiation. The proposed role for IGF-1 in the induction of oligodendrocyte differentiation and stimulation of myelination is in accordance with the synthesis of IGF-1 in selective cellular sites during their synaptogenesis and myelination. Experimentally induced demyelination induces IGF-1 synthesis in astrocytes in areas of demyelination and a transient increase in IGF-1R expression in regenerating oligodendrocytes (21). Neuronal and glial cell metabolism may also be regulated by IGFs. For example, Werner *et al.* (42) have demonstrated that IGF-1 increases expression of the brain glucose transporter.

IGF-1 has also been implicated in the modulation of neurotransmitter release. This action is due to the amino-terminal tripeptide released following post-translational procession of the IGF-1 precursor protein (26). GPE cross-reacts in one of the NMDA receptors for excitatory amino acids in the brain and via this receptor modulates dopamine release from neurons as well as neuronal excitability in response to glutamate (26). GPE also displays a potentiating action on the release of acetylcholine from neurons.

INSULIN-LIKE GROWTH FACTOR IN DOWN'S SYNDROME

Since DS is characterized by both mental and growth retardation and no GH deficiency has been found, it was of great interest to study the IGFs and the IGF receptors in DS. In the fetal circulation scarcely any radioreceptor assay (RRA)-IGF activity was found in fetuses with trisomy 21 (43). With the RRA-IGF assay all IGF activity is detected. From age two years patients with DS demonstrated a serum deficit in radioimmunoassay (RIA)-IGF-1 activity, which failed to increase during childhood and adolescence (Fig. 1) (2). By contrast, throughout postnatal life the serum levels of RRA-IGF were elevated (Fig. 1) (2). When pooled DS serum was acidified and separated by gel chromatography on Sephadex G-50 equilibrated with acetic acid, most RIA-IGF-1 and all RRA-IGF were dissociated from their carrier protein and recovered in the low molecular weight range (2). However, the serum levels of RIA-IGF-2 in DS patients were normal throughout life (Fig. 1) (44). Normal levels of receptors for IGFs and insulin were also found in DS (45).

Thus patients with DS seem to have a selective IGF-1 deficiency. The elevated serum level of RRA-IGF in DS patients seems not to be attributed to the presence of binding proteins, since the samples had been acidified before assay and subjected to acid gel chromatography. The finding could be explained by the presence of the fetal variant form of IGF-1 in the circulation long after birth. The result of our studies suggests that in DS there is a delayed maturation of the IGF system, with incomplete switching from production of the fetal form of IGF-1 to the GH-regulated IGF-1. It could also be explained by a partial GH deficiency, not recognized by present standard methods.

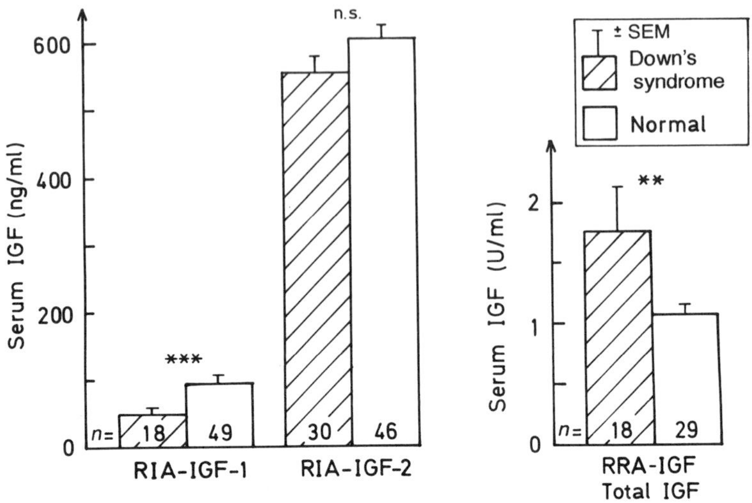

Figure 1. Mean serum levels (mean + 1 SEM) of IGF-1, (RIA-IGF-1; ng/ml), IGF-2 (RIA-IGF-2; ng/ml) and total IGF (RRA-IGF; U/ml) in children with Down's syndrome and healthy children. Number of children in each group (*n*) as well as levels of significance when the two groups of children are compared (not significant = n.s., $^{**}p < 0.01$, $^{***}p < 0.001$ are indicated. Reproduced with permission from *Am. J. Med. Genet.*, Suppl. 1990; **7:** 59–62

RESPONSE TO GROWTH HORMONE THERAPY IN CHILDREN WITH DOWN'S SYNDROME

Since DS individuals have shown a selective deficiency of the GH-dependent IGF-1 and since some healthy short children respond to human GH (hGH) therapy, it was of interest to investigate the effect of such treatment in DS children (5). Five children with DS—three girls and two boys of age three and a half to six years and with profound growth retardation and a mean height below −4 SD from the Swedish standard (46)—were chosen for six months' hGH treatment. None of them had clinical signs of increased susceptibility to infection or chronic diseases. Before hGH treatment was begun, the endogenous production of GH, as determined by nocturnal pulsatile secretion and after stimulation, as well as the serum levels of insulin, thyroid hormones and cortisol, were examined and found to be normal. The serum levels of IGF-1 were low before treatment in all five children.

During the six months of hGH therapy the serum values of both IGF-1 and IGF-2 increased to normal in all five children. The growth velocity increased in all the patients from a mean of 2.5 cm per six months before hGH treatment to 4.8 cm per six months during the six months of treatment. DS children show retardation of skeletal maturation and a dissociation between bone age and bone elongation for considerable periods (47). In the study the skeletal maturation was retarded in all patients, and did not accelerate as a result of the treatment (5).

The serum level of GH appeared to be normal in the children presented here, as well as in earlier studies (3, 4). Growth becomes GH dependent during the first year of life (48). This may be the critical period for imprinting of the GH–IGF system in man. We propose that there is a delay in the development of the IGF system in DS. This could result in abnormalities during the critical imprinting period, resulting in diminished sensitivity to GH (49).

GROWTH HORMONE THERAPY IN VERY YOUNG CHILDREN WITH DOWN'S SYNDROME: PRESENTATION OF AN ONGOING TRIAL

The aim of a present ongoing trial is to study the long-term effects of hGH therapy in young children with DS on linear growth as well as on brain growth and mental and motor development and on levels of IGFs in serum and CSF (50). The study group consists of 16 children (12 boys and four girls) with DS. The children have no congenital malformations of the heart. The children were between six and nine months of age (mean 7.4 months) at the start of hGH therapy. They are all cared for by their biological parents and none is malnourished.

Anthropometric measurements (height, head circumference and weight) were compared with 40 Swedish children with DS, without any congenital heart malformation. Growth was also compared with the standard growth charts for healthy Swedish children (46) and that of children with DS (1). Auxiological measurements as well as video recordings of neurological examinations were performed at the start of the trial, every third month during the first year and every six months during the second and third years of the trial as well as six and 12 months after the trial was completed. Since there were 12 boys and only four girls in the study group all the children were compared with growth charts for boys. Since boys are slightly taller than girls, the effects of hGH therapy might be somewhat underestimated. Bone age determinations according to Greulich and Pyle (GP) (51) and Tanner and Whitehouse (TW2) (52), as well as examinations of mental development (Griffiths' mental development scale) (53) and motor development (motor-perceptual development test) (54) and CSF sampling for IGF analyses, were performed at the start, after one year, at the end and one year after the end of the trial. The following blood analyses were performed at the start, every six months of the trial, and one year after the treatment period: thyroid hormones (thyrotropin, thyroxine and triiodothyronine), IGFs, routine hematology, bilirubin, liver enzymes (aminotransferases and alkaline phosphatase), creatinine and gliadin autoantibodies. Five days after the start of hGH therapy serum samples for

IGF analyses were taken. No results concerning mental or motor development, blood or CSF IGF levels have yet been evaluated.

TREATMENT

Three years of daily injections of recombinant hGH (Genotropin® , Kabi-Pharmacia, Stockholm, Sweden), 0.1 U/kg body weight per day from age six to nine months until three and a half to three and three quarter years of age.

ETHICAL EVALUATION

The trial was approved by the Ethics Committee of the University of Uppsala and by the Medical Products Agency of Sweden.

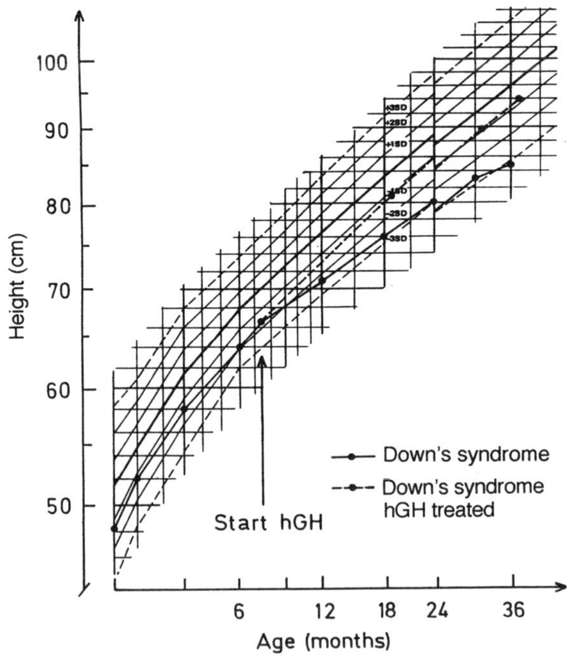

Figure 2. Mean height of children with Down's syndrome treated with growth hormone (hGH) for 30 months, from the age of six to nine months of age (mean 7.43 months, $n = 16$) compared to children with Down's syndrome without hGH therapy. The results are plotted against the standard growth chart for healthy Swedish boys. From *Acta Paediatr. Scand.* Suppl. 256, 1976. Reproduced by permission of Scandinavian University Press

PRESENT RESULTS FROM THE ONGOING HUMAN GROWTH HORMONE TREATMENT STUDY

In the present study we report the results of hGH therapy upon growth of DS children. The results after 12 ($n = 16$), 24 ($n = 12$) and 30 ($n = 8$) months are presented. None of the children had any other abnormality concerning blood chemistry during treatment, except for slightly elevated levels of liver enzymes. All of the children tolerated the injections well. Less than one injection per month of treatment has been neglected by the parents.

The mean height standard deviation score (SDS) (range; Swedish standard) before therapy was −1.8 (−0.5 to −3.1). After 12, 24 and 30 months the mean height SDS were −1.1 (−0.8 to −1.9), −0.9 (0 to −1.5) and −0.9 (0.1 to −1.5). During the first year of hGH treatment the children with DS gained height and then followed the growth rate of healthy Swedish children (46) (Fig. 2). As compared to growth charts for children with DS (1) the mean height of our children started at the 50th centile, reaching the 95th centile after 24 months of treatment (Fig. 3). The growth chart concerning height of the 40 control DS children did not differ from the standard curve of DS children from Cronk *et al.* (1) (Fig. 3). The mean bone age did not accelerate further during the first year of hGH therapy and was still retarded by about six months after one year of hGH therapy.

The mean head circumference was −1.2 (−0.4 to −3.5) before therapy. After 12, 24 and 30 months the mean head circumference SDS were −1.1 (0 to −2.5), −1.1 (0 to −2.2) and −1.2 (−0.5 to −2.0) respectively. The hGH

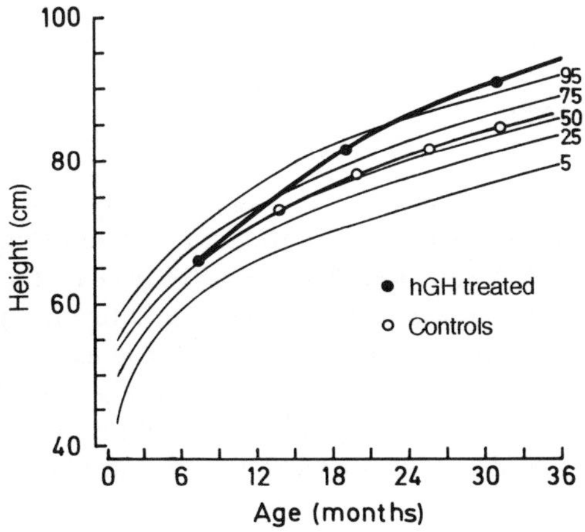

Figure 3. Mean height of children with Down's syndrome treated with human growth hormone (hGH) for 30 months, from the age of six to nine months (mean 7.43 months; $n = 16$), compared to standard growth chart for stature of boys with Down's syndrome. From *Pediatrics*, 1988; **81:** 102–110. Reproduced by permission of *Pediatrics*. The mean height of 40 Swedish children with DS, without congenital malformation of the heart, is also indicated

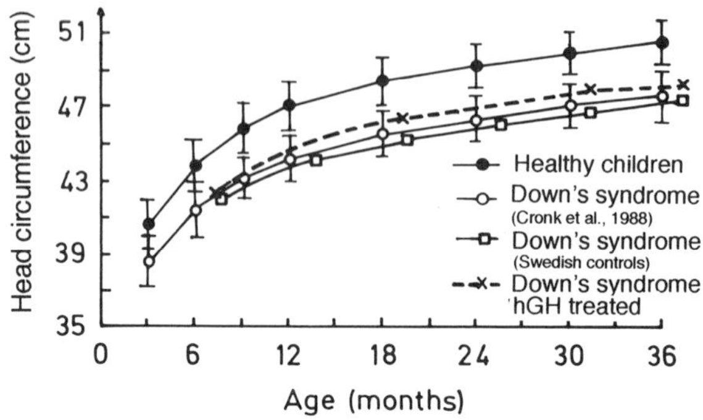

Figure 4. Mean head circumference of children with Down's syndrome treated with human growth hormone (hGH) for 30 months, from the age of six to nine months (mean 7.43 months; $n = 16$) compared to mean head circumference ($\pm$SD) for boys with Down's syndrome and to the standard from the National Center for Health Statistics (USA). From *Pediatrics*, 1988; **81:** 102–110. Reproduced by permission of *Pediatrics*. The mean head circumference of 40 Swedish children with DS, without congenital malformation of the heart, is also indicated

therapy resulted in only a slight increase of the growth velocity of the head circumference compared to the 40 Swedish control DS children or the growth chart of Cronk *et al.* (1) (Fig. 4). The growth in head circumference during hGH therapy seems not to correspond to the marked increase in longitudinal growth.

FINAL COMMENTS

The growth velocity in DS is most markedly reduced compared to healthy children from the age of six to nine months of age and until the age of three years (1, 2). During this age period the retardation in mental development becomes obvious and there is a decline in mean intelligence quotient (IQ) from IQ 70 at the age of nine months to IQ 40–50 at the age of three years. After this time there is only a slight further decline in IQ (55). Consequently, we chose to start GH therapy at the age of six to nine months in order to study whether GH therapy could prevent a further retardation of growth and mental development.

From the present ongoing study we can so far only report on physical growth. The markedly reduced growth velocity in children with DS between six months and three years of age seems to be avoided by hGH therapy. The children started at a mean height of −1.8 SD in accordance with Swedish standard (46) and did not deviate further during treatment. The strong effect on linear growth velocity is very obvious when growth during hGH treatment is compared with the standard growth chart of children with DS (1). All these GH-treated children were above the 95th centile for DS children after 24 months of therapy. During the first year of treatment no acceleration in bone age development was observed. Although the effect on linear growth velocity is obvious, it is too early to evaluate any long-term effect with regard to final height. There are so far no results available concerning skeletal maturation after three years of treatment. If it is still retarded, a taller final height would be predicted. However, growth velocity might become markedly reduced when GH treatment is withdrawn. Consequently, it is important to follow the children for long periods after hGH therapy has stopped. In two of the three children who so far have finished GH treatment, as well as in three out of five children who were treated eight years ago (5), the increased growth velocity continued during the following year. This might indicate a good long-term effect of hGH on linear growth in DS. It might also indicate that a short-term period of hGH treatment might result in a maturation of the IGF–GH system.

This is an ongoing study and results concerning the mental or motor development or concerning IGF levels have not yet been assessed. There is, however, a very slight increase in head circumference observed, which could be attributed to the hGH therapy. This is not in accordance with the finding of Torrado *et al.* (56), who reported a significant increase in growth velocity of head circumferences. Our result might indicate that the microcephaly in DS is only partly a result of growth retardation and is mainly due to reduced brain growth, which is not improved by GH treatment. Because of this, it might be questioned whether an improvement in mental development in DS could be achieved by hGH treatment. The growth of the CNS has never been reported to be related to GH, although IGFs seem to have effects on the growth of the CNS. Furthermore, GH-deficient dwarfs are not mentally retarded. On the other hand, GH-deficient adults are reported to have beneficial mental effects from hGH therapy, with improved psychological capacity (57). Our study will help us understand the growth retardation in DS and indicate whether or not GH has any impact on mental development. Until more knowledge concerning the effects of GH therapy in DS is obtained GH treatment cannot be recommended in clinical practice to children with DS.

ACKNOWLEDGEMENTS

This study was supported by the First of May-flower Foundation, the Sven Jerring Foundation, the Loo and Hans Osterman Foundation, Kabi-Pharmacia Peptide Hormones, Stockholm, Sweden, the Sävstaholm Foundation, the Swedish Medical Research Council (grants nos 5445, 6446 and 2604) and the Swedish Cancer Foundation.

REFERENCES

1. Cronk C, Crocker AC, Peuschel SM, Shea AM, Zackai E, Pickens G, Reed RB. Growth charts for children with Down's syndrome: 1 month to 18 years of age. *Pediatrics* 1988; **81:** 102–110.
2. Sara VR, Gustavson K-H, Annerén G, Hall K, Wetterberg L. Somatomedins in Down's syndrome. *Biol Psychiatry* 1983; **18:** 803–811.
3. Milunsky A, Lowy C, Rubenstein A, Wright AD. Carbohydrate tolerance, growth hormone and insulin levels in mongolism. *Dev Med Child Neurol* 1968; **10:** 25–31.
4. Ruvalcaba RHA, Thuline HC, Kelley VC. Plasma growth hormone in patients with chromosomal anomalies. *Arch Dis Childh* 1972; **47:** 307–309.
5. Annerén G, Sara VR, Hall K, Tuvemo T. Growth and somatomedin responses to growth hormone in Down's syndrome. *Arch Dis Childh* 1986; **61:** 48–52.
6. Castells S, Torrado C, Bastian W, Wisniewski KE. Growth hormone deficiency in Down's syndrome. *J Intell Disab Res* 1992; **36:** 29–43.
7. Ayer-LeLievre C, Ståhlbom P-A, Sara VR. Expression of IGF-1 and IGF-2 mRNA in the brain and cranio-facial region of the rat fetus. *Dev Neurol* 1991; **111:** 105–115.
8. Bondy CA. Transient IGF-1 gene expression during the maturation of functionally related central projection neurons. *J Neurosci* 1991; **11:** 3442–3455.
9. Bondy CA, Werner H, Roberts CT Jr, LeRoith D. Cellular pattern of insulin-like growth factor-I (IGF-I) and type 1 IGF receptor gene expression in early organogenesis: comparison with IGF-II gene expression. *Mol Endocrinol* 1990; **4:** 1386–1398.
10. Werther GA, Abate M, Hogg A, Cheesman H, Oldfield B, Hards D, Hudson P, Power B, Freed K, Herington AC. Localisation of insulin-like growth factor mRNA in rat brain by in situ hybridization: relationship to IGF-I receptors. *Mol Endocrinol* 1990; **4:** 773–778.
11. Wood TL, Brown AL, Rechler MM, Pintar JE. The expression pattern of an insulin-like growth factor (IGF)-binding protein gene is distinct from IGF-II in the midgestational rat embryo. *Mol Endocrinol* 1990; **4:** 1257–1263.
12. Lamson G, Pham H, Oh Y, Ocrant I, Scwander J, Rosenfeld RG. Expression of the BRL-3A insulin-like growth factor binding protein (rBP-30) in the rat central nervous system. *Endocrinology* 1989; **123:** 1100–1102.
13. Roghani M, Lassarre C, Zapf J, Povoa G, Binoux M. Two insulin-like growth factor (IGF)-binding proteins are responsible for the selective affinity for IGF-II of cerebrospinal fluid binding proteins. *J Clin Endocrinol Metab* 1991; **73:** 658–666.
14. Tseng L Y-H, Brown AL, Yang Y W-H, Romanus JA, Orlowski CC, Taylor T, Rechler MM. The fetal rat binding protein for insulin-like growth factors is expressed in the choroid plexus and cerebrospinal fluid of adult rats. *Mol Endocrinol* 1989; **3:** 1559–1568.
15. Roth RA, Steele-Perkins G, Hari J, Stover C, Pierce S, Turner J, Edman JC, Rutter WJ. Insulin and insulin-like growth factor receptors and responses. *Cold Spring Harbor Symp Quant Biol* 1988; **LIII:** 537–543.
16. Sara VR, Hall K, Mizaki M, Fryklund L, Christensen N, Wetterberg L. The ontogenesis of somatomedin and insulin receptors on the human fetus. *J Clin Invest* 1983; **71:** 1084–1094.
17. Baron-Van Evercooren A, Olichon-Berthe C, Kowalski A, Visciano G, Van Obberghen E. Expression of IGF-1 and insulin receptor genes in the rat central

nervous system: a developmental, regional, and cellular analysis. *J Neurosci Res* 1991; **28:** 244–253.
18. Bondy CA, Werner H, Roberts CT Jr, LeRoith D. Cellular pattern of type-I insulin-like growth factor receptor gene expression during maturation of the rat brain: comparison with the insulin-like growth factors I and II. *Neuroscience* 1992; **46:** 909–923.
19. Valentino KL, Ocrant I, Rosenfeld RG. Developmental expression of insulin-like growth factor II receptor immunoreactivity in the rat central nervous system. *Endocrinology* 1990; **126:** 914–920.
20. Kovacina KS, Steele-Perkins G, Roth RA. A role for the insulin-like growth factor II/mannose-6-phosphate receptor in the insulin-induced inhibition of protein catabolism. *Mol Endocrinol* 1989; **3:** 901–906.
21. Komoly S, Hudson LD, Webster HD, Bondy CA. Insulin-like growth factor I gene expression is induced in astrocytes during experimental demyelination. *Proc Natl Acad Sci USA* 1992; **89:** 1894–1898.
22. Lee W-H, Clemens JA, Bondy CA. Insulin-like growth factors in the response to cerebral ischemia. *Mol Endocrinol* 1992; **3:** 36–43.
23. Sandberg-Nordqvist A-C, Ståhlbom P-A, Lake M, Sara VR. Characterization of two cDNAs encoding insulin-like growth factor 1 (IGF-1) in the human fetal brain. *Mol Brain Res* 1992; **12:** 275–277.
24. Sara VR, Carlsson-Skwirut C, Andersson C, Hall K, Sjögren B, Holmgren A, Jörnvall H. Characterization of somatomedins from human fetal brain: identification of a variant of insulin-like growth factor 1. *Proc Natl Acad Sci USA* 1986; **83:** 4904–4907.
25. Carlsson-Skwirut C, Lake M, Hartmanis M, Hall K, Sara VR. A comparison of the biological activities of recombinant intact and truncated insulin-like growth factor 1 (IGF-1). *Biochim Biophys Acta* 1989; **1011:** 192–197.
26. Sara VR, Carlsson-Skwirut C, Bergman T, Jörnvall H, Roberts PJ, Crawford M, Nilsson-Håkansson L, Civalero I, Nordberg A. Identification of Gly-Pro-Glu (GPE), the aminoterminal tripeptide of insulin-like growth factor 1, which is truncated in brain, as a novel neuroactive peptide. *Biochem Biophys Res Commun* 1989; **165:** 766–771.
27. Sandberg AC, Engberg C, Lake M, von Holst H, Sara VR. The expression of IGF-1 and IGF-2 genes in the human fetal and adult brain and in glioma. *Neurosci Lett* 1988; **93:** 114–119.
28. Bach MA, Shen-Orr Z, Lowe WL Jr, Roberts CT Jr, LeRoith D. Insulin-like growth factor 1 mRNA levels are developmentally regulated in specific regions of the rat brain. *Mol Brain Res* 1991; **10:** 433–448.
29. Hepler JE, Lund PK. Molecular biology of the insulin-like growth factors: relevance to nervous system function. In *Molecular Neurobiology*, edited by Bazan N, Human Press, Clifton, USA, 1990; pp. 93–127.
30. Phillipps A, Drakenberg K, Persson B, Sjögren B, Eklöf AC, Hall K, Sara VR. The effects of altered nutritional status upon insulin-like growth factor synthesis in neonatal rats. *Pediatr Res* 1989; **26:** 128–134.
31. Strauss DS, Takemoto CD. Specific decrease in liver insulin-like growth factor-I and brain insulin-like growth factor-II gene expression in energy-restricted rats. *J Nutr* 1991; **121:** 1279–1286.
32. Adamo M, Werner H, Farnsworth W, Roberts CT Jr, Raizada M, LeRoith D. Dexamethasone reduces steady state insulin-like growth factor I messenger ribonucleic acid levels in rat neuronal and glial cells in primary culture. *Endocrinology* 1988; **123:** 2565–2570.
33. Drago J, Murphy M, Carroll SM, Harvey RP, Bartlett P. Fibroblast growth

factor-mediated proliferation of central nervous system precursors depends on endogenous production of insulin-like growth factor I. *Proc Natl Acad Sci USA* 1991; **88:** 2199–2203.
34. Torres-Aleman I, Naftolin F, Robbins RJ. Trophic effects of basic fibroblast growth factor on fetal rat hypothalamic cells: interaction with insulin-like growth factor I. *Dev Brain Res* 1990; **52:** 253–257.
35. DiCicco-Bloom E, Black IB. Insulin-like growth factors regulate the mitotic cycle in cultured rat sympathetic neuroblasts. *Proc Natl Acad Sci USA* 1988; **85:** 4066–4070.
36. Giacobini M-B, Olson L, Sara VR. A truncated form of IGF-1 found in brain exerts trophic effects on fetal brain tissue grafts. *Exp Neurol* 1990; **108:** 33–37.
37. McMorris FA, Smith TM, DeSalvo S, Furlanetto RW. Insulin-like growth factor I/somatomedin C: a potent inducer of oligodendrocyte development. *Proc Natl Acad Sci USA* 1986; **83:** 822–826.
38. Nielsen FC, Wang E, Gammeltoft S. Receptor binding, endocytosis, and mitogenesis of insulin-like growth factor I and II in fetal rat brain neurons. *J Neurochem* 1991; **56:** 12–21.
39. Påhlman S, Meyerson G, Lindgren E, Schalling M, Johansson I. Insulin-like growth factor I shifts from promoting cell division to potentiating maturation during neuronal differentiation. *Proc Natl Acad Sci USA* 1991; **88:** 9994–9998.
40. Recio-Pinto E, Lang FF, Ishii DN. Insulin and insulin-like growth factor binding and the neurite formation response in cultured human neuroblastoma cells. *Proc Natl Acad Sci USA* 1984; **81:** 2562–2566.
41. Recio-Pinto E, Rechler MM, Ishii DN. Effects of insulin, insulin-like growth factor-II, and nerve growth factor on neurite formation and survival in cultured sympathetic and sensory neurons. *J Neurosci* 1986; **6:** 1211–1219.
42. Werner H, Raizada MK, Mudd LM, Foyt HL, Simpson IA, Roberts CT Jr, LeRoith D. Regulation of rat brain/HepG2 glucose transporter gene expression by insulin and insulin-like growth factor-I in primary cultures of neuronal and glial cells. *Endocrinology* 1989; **125:** 314–320.
43. Sara VR, Hall K, Rodeck CH, Wetterberg L. Human embryonic somatomedin. *Proc Natl Acad Sci USA* 1981; **78:** 3175–3179.
44. Annerén G, Engberg G, Sara VR. The presence of normal levels of serum immunoreactive insulin-like growth factor 2 (IGF-2) in patients with Down's syndrome. *Ups J Med Sci* 1984; **89:** 274–278.
45. Sara, VR, Sjögren B, Annerén G, Gustavson K-H, Forsman A, Hall, K, Wahlström J, Wetterberg L. The presence of normal receptors for somatomedin and insulin in fetuses with Down's syndrome. *Biol Psychiatry* 1984; **19:** 591–598.
46. Karlberg P, Taranger J, Engström J, Landström T, Lichtenstein, H, Lindström B, Svennberg-Redegren I. Physical growth from birth to 16 years and longitudinal outcome of the study during the same period. *Acta Paediatr Scand* 1976; Suppl. 258.
47. Roche AF. The stature of mongols. *J Ment Defic Res* 1963; **9:** 131–145.
48. Root AW. Growth hormone and prolactin in the fetus. *Prog Clin Biol Res* 1976; **10:** 107–126.
49. Annerén G, Gustavson K-H, Sara VR, Tuvemo T. Growth retardation in Down's syndrome in relation to insulin-like growth factors and growth hormone. *Am J Med Genet* 1990; (Suppl.) **7:** 59–62.
50. Annerén G, Gustafsson J, Sara VR, Tuvemo T. Normalized growth velocity in children with Down's syndrome during growth hormone therapy. *Intell Disab Res* 1993, in press.

51. Greulich WW, Pyle SJ. *Radiographic Atlas of the Skeletal Development of Hand and Wrist*. Stanford University Press, Stanford, 1959.
52. Tanner M, Whitehouse RH. Clinical longitudinal standards for height, weight, height velocity and the stages of puberty. *Arch Dis Childh* 1980; **19:** 790–797.
53. Alin-Akerman B, Nordberg L. *Griffiths' utvecklingskalor I och II. Manual för administrering och utvärdering*. Psykologi-förlaget AB, Eskilstuna, Sweden, 1980.
54. Holle B, Bonnelycke K, Kemp E, Thrane-Mortensen L. *Motorisk-Perceptuell utveckling. Protokoll 0-7 år*. Psykologi-förlaget AB, Eskilstuna, Sweden, 1984.
55. Carr J. Six weeks to twenty-one years old: longitudinal study of children with Down's syndrome and their families. *J Child Psychol Psychiatry* 1988; **29:** 407–431.
56. Torrado C, Bastian W, Wisniewski KE, Castells S. Treatment of children with Down's syndrome and growth retardation with recombinant human growth hormone. *J Pediatr* 1991; **119:** 478–483.
57. McGauley GA. Quality of life assessment before and after growth hormone treatment in adults with growth hormone deficiency. *Acta Paediatr Scand* 1989; (Suppl.) **356:** 70–72.

Discussion of Presentation

GÖRAN ANNERÉN
(Edited by D.A. Snider and Salvador Castells)

C. Torrado: How were you able to assess patient compliance? How did you know the parents were indeed giving the GH to the children?

Z. Laron: You measured serum IGF-1 concentrations and could use that as an index of compliance.

G. Annerén: Yes, and in all patients the serum IGF-1 levels increased.

S. Castells: We follow IGF-1 levels, too, and the levels stayed up. If the patients stop taking the injections of hGH, the serum IGF-1 concentrations return to the levels before treatment. Serum IGF-1 concentrations are a very good index of compliance. We only treated patients who had a lack of response to clonidine and L-dopa. I see that in your testing you used insulin and arginine, which are stimuli to the hypothalamus and pituitary. Some of the amino acids appear to stimulate the pituitary directly. Arginine could then be considered as a direct stimulus to the pituitary; insulin-induced hypoglycemia is known to be a hypothalamic stimulus. By using both together, you stimulate the pituitary and the hypothalamus at the same time and have difficulty in differentiating and isolating hypothalamic dysfunction. In our case, we use L-dopa and clonidine, drugs known to stimulate hypothalamic centers only and by stimulating GHRH to increase serum concentration of hGH. We had a few children who had normal responses to both tests, and we did not treat with hGH. We feel that about 5% of children with Down's syndrome grow normally and have normal responses to L-dopa and clonidine. In your case, you seem to have treated these children with recombinant hGH. There also appears to be a difference in the amount of GH deficiency among Down's syndrome children.

G. Annerén: We have stopped using the insulin tests and are doing only the sleeping curve for 8 h. We repeat this test one year after the treatment is finished. Concerning the mental effects, the parents are very happy with the treatment, and it is very difficult to say there is no effect. There could be a placebo effect. That is why it is very important to test the performance. A control group seems to do exactly the same as the recombinant hGH treated children. The control group is following a medical care program and a special communication training program, but without any hGH therapy.

15 Preliminary Results after One Year of Recombinant Human Growth Hormone Treatment in Four Subjects with Down's Syndrome

LETIZIA RAGUSA,[1] ANTONINO ALBERTI,[1] CATERINA PROTO,[2] CORRADO ROMANO[1*] AND FABIO COLABUCCI[1,3]

[1]*Department of Pediatrics and* [2]*Radioimmunology Service, OASI Institute, Troina, and* [3]*Department of Pediatrics, Catholic University, Rome, Italy*

INTRODUCTION

Growth retardation is a characteristic feature of Down's syndrome (DS) as it has been documented both in the prenatal and postnatal periods. Annerén *et al.* (1) found low serum concentrations of insulin-like growth factor I (IGF-1), thus explaining the growth retardation as a delayed maturation and incomplete switching from production of the fetal to the growth hormone (GH)-regulated IGF-1. Torrado *et al.* (2) observed 13 GH-deficient DS patients. We (3) stressed the low 24 h integrated concentration of GH (ICGH/24 h) found in two DS children. Annerén *et al.* (4) treated with recombinant human GH (rhGH) five DS subjects for six months, achieving a doubling of the growth rate. The 13 patients of Torrado *et al.* (2) were given rhGH for one year, with a significant increase in annual growth rate and head circumference. Here we present our results on rhGH treatment in four DS subjects with low ICGH/24 h.

MATERIALS AND METHODS

We evaluated GH secretion and IGF-1 serum levels in 15 DS patients (nine males and six females), aged 6.5–17.3 years.

*Address for Correspondence: Department of Pediatrics, OASI Institute (IRCCS), Via Conte Ruggero, 73, 94018 Troina (EN), Italy.

Growth Hormone Treatment in Down's Syndrome. Edited by S. Castells and K.E. Wisniewski

Subject no.	pv CT (ng/ml)	pv ITT (ng/ml)	ICGH/24 h (ng/ml)	IGF-1 (NV) (IU/ml)
1	25.41	6.47	2.09	0.43 (0.99–6.80)
2	8.71	9.45	1.62	0.20 (0.28–3.71)
3	8.68	19.39	1.65	0.16 (0.28–3.71)
4	1.66	6.24	1.32	0.21 (0.22–2.80)
5	5.56	41.83		0.9 (0.71–4.10)
6	21.60	12.33	3.55	1.06 (0.90–5.60)
7	7.83	22.70		0.84 (0.99–6.80)
8	25.25	38.92		0.67 (0.22–2.80)
9	4.07	40.47		1.15 (1.20–5.90)
10	6.22	18.59		0.62 (0.28–3.07)
11	9.95	39.11		0.61 (0.28–3.07)
12	11.26	13.53	3.53	0.81 (0.90–5.60)
13	22.94	24.46		0.96 (0.91–3.11)
14	15.07	9.26		0.73 (0.99–6.80)
15	7.98	17.41		0.65 (0.99–6.80)

pv = peak value

Figure 1. GH assessment in 15 DS subjects

Figure 1 shows the GH assessment in our subjects performed according to the following work-up:

1. Insulin tolerance test (ITT).
2. Clonidine test (CT).
3. ICGH/24 h.
4. IGF-1 serum levels.

Subjects 1, 2, 3 and 4 with low IGF-1 serum levels, peak serum GH concentrations < 10 ng/ml after ITT and/or CT, and ICGH/24 h < 3.2 ng/ml were given rhGH (0.8 U/kg per week) for one year.

RESULTS

Figure 2 shows the height growth rate (HGR) in the four subjects treated with rhGH. We compared the group of four rhGH-treated DS subjects (1F : 3M, mean age 10.1 ± 3.3 years) with a control group of six DS subjects (2F : 4M, mean age 10.6 ± 1.9 years).

Figure 3 shows the statistical analysis of HGR in both groups.

DISCUSSION

Recombinant hGH therapy seems to be effective in increasing the HGR of DS children. The results reported by Annerén *et al.* (4) and Torrado *et al.* (2) are confirmed in the present study. Furthermore, we compared the HGR in

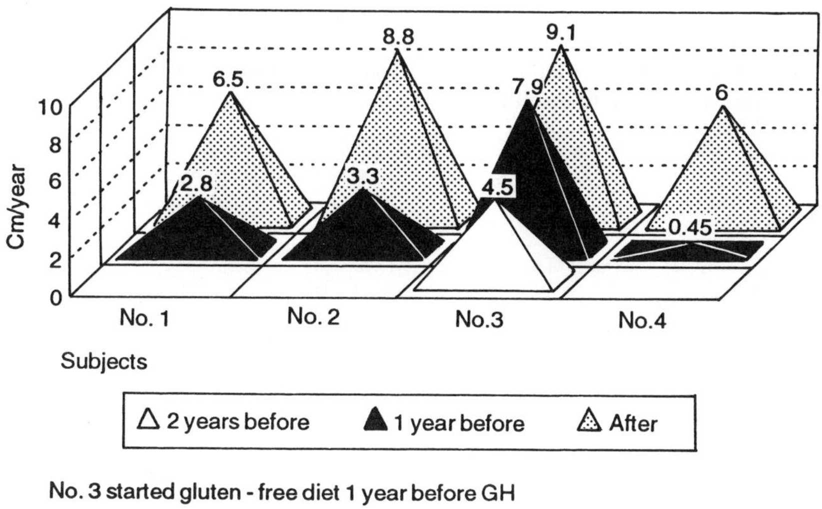

Figure 2. Height growth rate before and after GH treatment

	rhGH DS (cm/year	control DS (cm/year	Mann–Whitney test
1st year	3.61 ± 3.11	4.21 ± 2.03	n.s.
2nd year	7.60 ± 1.57	3.60 ± 1.23	$p = 0.005$

In the first group 2nd year indicates the year of rhGH treatment.

Figure 3. Statistical analysis of growth rate

rhGH-treated and untreated DS children over two years. No significant differences were shown without therapy (first year). In the second year the first group was given rhGH, with a significant ($p = 0.005$) increase in HGR, in comparison with the control group. We would like also to suggest the GH evaluation in DS within a comprehensive work-up, where the possibility of celiac disease must be screened. Indeed, subject 3 in Fig. 2 shows that a gluten-free diet is able to increase the HGR in DS children.

REFERENCES

1. Annerén G, Sara VR, Hall K, Tuvemo T. Growth and somatomedin responses to growth hormone in Down's syndrome. *Arch Dis Child* 1986; **61:** 48–52.
2. Torrado C, Bastian W, Wisniewski KE, Castells S. Treatment of children with Down's syndrome and growth retardation with recombinant human growth hormone. *J Pediatr* 1991; **119**(3): 478–483

3. Ragusa L, Romano C, Failla P, Proto C, Colabucci F. Growth hormone subnormality in Down's syndrome. *Am J Med Genet* 1992; **43:** 894–895.
4. Annerén G, Gustavson KH, Sara VR, Tuvemo T. Growth retardation in Down's syndrome in relation to insulin-like growth factors and growth hormone. *Am J Med Genet* 1990; (Suppl.) **7:** 59–69.

16 Psychoendocrine–Immune Interactions in Down's Syndrome: Role of Zinc

NICOLA FABRIS, EUGENIO MOCCHEGIANI* AND GIORGIO ALBERTINI[1]

Department of Geriatrics and Gerontology, Italian National Research Center on Aging (INRCA), Ancona, and [1]Bambino Gesù Hospital, S. Marinella, Italy

INTRODUCTION

Down's syndrome (DS) is the most frequent chromosomal aberration in humans, characterized, in most cases, by an extra chromosome 21 (trisomy 21) (1). In addition to mental retardation, DS subjects have an increased cancer incidence and show signs of precocious aging of various organs and tissues (2). The disease is a complex clinical syndrome, where cardiopathies, high incidence of leukemia and increased susceptibility to infections are frequent causes of death (3).

From a physiopathological point of view, most of the homeostatic networks are found to be altered in DS subjects, namely the immune (4–18), neuroendocrine (19–27) and nervous (28–35) systems and the complex network regulating the turnover of nutrients, including vitamins and trace elements (36–47). The causes of such widespread alterations are still unknown, though some unifying hypotheses have recently been proposed. In particular, the existence of strict interrelationships among the major homeostatic systems, i.e. the nervous, neuroendocrine and immune networks (48), and the role played in these networks by a trace element, namely zinc (49), together with the findings of reduced turnover of this element in DS subjects, have led to the hypothesis that such a mineral may, at least partially, be involved in some of the physiopathological features of DS.

In the present chapter, the most relevant alterations in the homeostatic networks present in DS subjects will be reviewed and data will be reported on the beneficial effect of zinc supplementation on neuroendocrine–immune interactions in these subjects. A new protocol for an integrated and interdisciplinary analysis of the role of zinc in DS will be illustrated.

*Address for correspondence: Immunology Center, Department of Geriatrics and Gerontology, INRCA, Via Birarelli 8, 60100 Ancona, Italy.

Growth Hormone Treatment in Down's Syndrome. Edited by S. Castells and K.E. Wisniewski

ALTERATIONS IN THE HOMEOSTATIC NETWORK IN DOWN'S SYNDROME

IMMUNOLOGICAL ABNORMALITIES

It is well documented that DS is associated with increased susceptibility to infectious diseases, particularly in the respiratory and gastrointestinal tracts (50, 51) and at a cutaneous level (52).

The numerous immunological analyses performed have suggested that the major defects in this homeostatic domain are the following (Table 1):

1. A marked decrease in the absolute number of $CD3^+$ (total T-cells) and $CD4^+$ (T-helper/inducer cells), with a significant increment in $CD8^+$

Table 1. Immunological alterations in DS

Element	High	Low	References
T-cells			
CD3 (T-total)		+ +	(6, 8)
CD4 (T-helper)		+ +	(6, 8)
CD8 (T-suppressor)	+ +		(6, 8)
CD25 (IL-2 receptor)		+ +	(9)
PHA response		+ +	(7, 15)
ConA response		+	(15)
NK cells			
$NKH\text{-}1^+$ ($CD56^+$)	+ +		(6, 8)
NK basal activity (K562)		+ +	(10)
Polymorphonuclear cells			
Chemotaxis		+ +	(11)
Phagocytosis		+ +	(12)
Bactericidal activity		+ +	(12)
Cytokines			
IL-2 production		+ +	(9)
IL-1 production	+		(13)
γ-IFN production		+ +	(10)
Adhesion molecules			
LFA-1[a]	+ +		(14)
Immunoglobulins			
IgM		+ +	(15)
IgA, IgE, IgD, IgG	+		(16, 17)
IgG_1, IgG_3	+ + +		(18)
IgG_2, IgG_4		+ + +	(18)
C3, C4	+ +		(6)

[a]LFA-1 = lymphocyte functional antigen 1.

(T-suppressor/cytotoxic cells) and $CD56^+$ (non-MHC restricted cytotoxic cells) (7, 8).
2. A reduced capacity of peripheral blood lymphocytes from DS subjects to proliferate after mitogenic stimulation (phytohaemagglutinin (PHA) and concanavalin A (Con A)), particularly in the second decade of life (15); also skin reactivity to tuberculin is reduced (53).
3. A marked reduction of the cytotoxic capacity of DS natural killer (NK) cells in spite of the increase in $CD56^+$ cell number (10).
4. A strong reduction of both interleukin 2 (IL-2) receptor expression and IL-2 production in lymphocyte cultures from DS subjects (9); such a defect does not seem to depend on an intrinsic alteration of T-cells but rather on a defective trans-membrane transduction (54).
5. With regard to serum immunoglobulin levels, although some findings are contradictory (4), in general, immunoglobulin M (IgM) levels are found to be lower in DS, whereas IgG and IgA are found to be higher in DS than in normal subjects (15–17).
6. DS subjects show abnormally low levels of 'natural' antibodies (55) and reduced antibody production to various antigens, including influenza virus (56). DS subjects also show a compromised polymorphonuclear cell function (11, 12).

THYMIC ABNORMALITIES

The majority of the immune alterations observed in DS subjects seem to depend on defective thymic function. At this level, in fact, numerous alterations have been documented (Table 2). In particular:

1. Young DS subjects show quite a rapid thymic involution (30% more than

Table 2. Thymic alterations in DS

	High	Low	References
Thymic cells			
$CD1^+$		+	(58)
$CD3^+$		+	(58)
$CD4^+$		+	(58)
$CD8^+$		+	(58)
α/β TCR		+ +	(59)
Hassal's corpuscles	+ +		(57)
Epithelial cells		+	(57)
Thymic humoral factors			
Thymulin (active ZnFTS)		+ +	(36)
Thymulin (total FTS)	Normal		(36)

TCR = T-cell receptor.

physiological) (57), with reduction of the thymocyte pool able to differentiate into functionally mature T-cells (58); also α/β T-cell receptor (TCR) expression on thymocytes from DS subjects is reduced (59).

2. Thymic endocrine activity is strongly depressed in DS subjects (36). The plasma level of one of the best-known thymic hormones, the facteur timique serique (FTS), called thymulin in its zinc-bound active form (Zn-FTS), is greatly decreased in DS (36) (Table 2). According to findings from our laboratory, such a reduction is due to a defect in zinc, which is required to confer biological activity to thymulin (63). In DS, the thymus produces normal amounts of the hormonal peptide but is inactive because it is not bound to zinc.

Table 3. Neuroendocrine and neurotransmitters alterations in DS

Element	High	Low	References
Neuroendocrine			
GH		+ +	(19, 20)
Prolactin M	+ + +		(25)
F	+ +		(25)
FSH M	+ + +		(27)
F	+ + +		(27)
LH M	+ + +		(27)
F	+ + +		(27)
TSH	+ + +		(21–24)
T_4		+	(23)
T_3		+	(25)
rT_3		+ +	(21, 22)
IGF-1		+ +	(19, 20)
Insulin		+ +	(26)
Estradiol M	+ +		(25)
Neurotransmitters			
Opioid peptides	+ +		(28)
Acetylcholine		+ +	(29)
Serotonin		+ +	(30, 33)
Norepinephrine:			
Brain tissue		+ + +	(33)
CSF	+ +		(34)
Somatostatin		+ +	(32)
Vasopressin		+ +	(35)
Dopamine	+		(31)
Choline		+ + +	(33)
Biopterin	+ +		(34)

5-HT = 5-hydroxy-tryptamine; CSF = cerebral spinal fluid; IGF-1 = insulin-like growth factor I; F = female; M = male.

NEUROENDOCRINE AND NEUROTRANSMITTER ALTERATIONS

The main alterations in the neuroendocrine system and neurotransmitter network present in DS are the following (Table 3):

1. With regard to pituitary hormones, there is a deficit of growth hormone (GH) and insulin-like growth factor 1 (IGF-1) in DS subjects (19, 20, 61), while increased secretion of prolactin (25), of gonadotropins (27) and of thyrotropin is observed (21, 24).
2. With regard to peripheral endocrine functions, alterations have been described in thyroid hormone activity in DS subjects, with reduced serum levels of reverse triiodothyronine (rT_3) (21, 22) and reduced insulin secretion (26).
3. Neurotransmitter production, with few exceptions, is generally reduced in DS subjects: the cholinergic, serotoninergic and opioid systems are all altered in DS subjects (28–35) and such alterations may be involved in the defective development of brain tissue (62, 63).

NUTRITIONAL ALTERATIONS

Studies performed in the last decade have indicated the relevance of nutritional status in DS subjects, particularly during infancy. Although the causes are still unknown, DS subjects suffer from altered absorption of various trace elements and vitamins (Table 4). Plasma zinc, selenium and iron are low (36–40), whereas plasma copper concentrations are higher than in normal children (38).

Table 4. Nutritional alterations (trace elements and vitamins) in DS

Element or vitamin	High	Low	References
Zinc		+ +	(36–40)
Copper	+ +		(38)
Selenium		+ +	(37)
Calcium	+		(37)
Iron		+ +	(37)
Magnesium		+ +	(37)
Manganese		+ +	(37)
Aluminium	+ +		(41)
Lithium		+	(39)
Lead		+	(39)
Gallium		+ +	(42)
Taurine	+ +		(43)
Vit. B_6		+ +	(44)
Vit. B_{12}		+	(45)
Vit. C		+ +	(46)
Vit. E		+	(47)

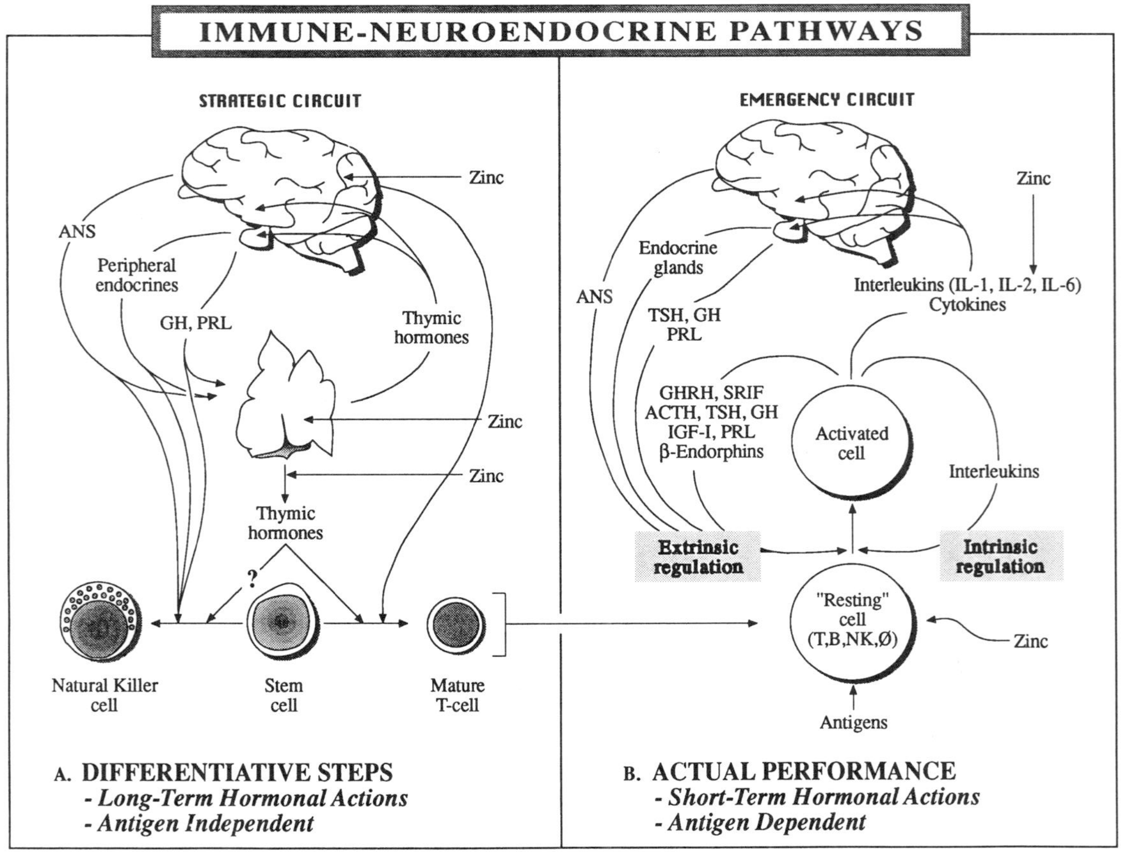

Figure 1. Schematic representation of the two major pathways of neuroendocrine–immune interactions: A (strategic circuit), B (emergency circuit). Zinc plays a crucial role at the level of the nervous, neuroendocrine and immune systems. (ANS = autonomous nervous system, GH = growth hormone, PRL = prolactin, TSH = thyrotropic stimulating hormone, GHRH = growth hormone releasing hormone, ACTH = adrenocorticotrophic hormone, SRIF = somatotrophic releasing inhibitory factor, IGF-1 = Insulin-like growth

A marginal zinc deficiency is particularly common in DS (36, 40): both plasma and lymphocyte zinc content are strongly reduced in DS (40). The alteration in the bioavailability of this trace element may play an important role in the pathogenesis of defects at the level of immune-neuroendocrine interactions.

ROLE OF ZINC IN IMMUNE–NEUROENDOCRINE INTERACTIONS

It is now accepted that the three major homeostatic mechanisms, i.e. the nervous, neuroendocrine and immune systems, are strictly interconnected (48). Evidence exists showing that a great number of circuits link these homeostatic systems both at a central level (strategic circuit), based on the interactions between the neuroendocrine system and the thymus, and at a peripheral level (emergency circuit) between neuroendocrine signals and the humoral products which are secreted by immune cells during specific reactions to various antigens (Fig. 1). The rationale for discriminating these two levels is based on various considerations. The first level of interaction is primarily involved in maturative steps, independent of the level of antigenic stimulation; the second level of interaction, on the contrary, requires the presence of fully differentiated immune cells and the occurrence of a specific antigenic or stress-mediated hormonal stimulus. The main role played by these interactions appears to be that of normalizing the neuroendocrine or immune balance when suddenly altered by a stressful cognitive or non-cognitive event. Zinc plays a crucial role for the efficiency of the three major systems of homeostasis and for their physiological interactions.

This trace element is present in high concentrations in discrete brain areas (64), it is relevant for the maintenance of a physiological hormonal profile (65) and is required to confer biological activity to thymulin (60), to nerve growth factor (NGF) (66) and to some immune cytokines, such as IL-1 and IL-6 (67). Finally, no less than 200 enzymes require zinc for their efficient action (65). We have shown that a marginal, but biologically relevant, zinc deficiency is usually present in children with DS and low bioavailability of zinc may play an important role in the pathogenesis of some immune and hormonal defects associated with the syndrome. This assumption is supported by a large quantitative amount of clinical evidence, which is summarized here.

EFFECT OF ZINC SUPPLEMENTATION IN DOWN'S SYNDROME SUBJECTS

Zinc supplementation (1 mg Zn^{2+}/kg body weight per day) for two to four months per year has been given to groups of DS subjects for three to four years, and various immunological and hormonal functions were analyzed

Table 5. Effect of oral zinc supplementation on nutritional and immune parameters in DS

		Cycle I			Cycle II		
Parameter	Source	Before	After	p	Before	After	p
Zinc (μg/dl)	P	85.8 ± 4.5	126.9 ± 9.2	<0.01	71.3 ± 3.4	100.2 ± 5.3	<0.01
Copper (μg/dl)	P	74.8 ± 3.8	79.6 ± 2.8	n.s.	67.9 ± 2.3	65.0 ± 2.8	n.s.
Active thymulin ($\log_{-2}$)	P	1.1 ± 0.2	4.3 ± 0.3	<0.01	2.1 ± 0.3	4.2 ± 0.2	<0.01
Total thymulin ($\log_{-2}$)	P	5.0 ± 0.3	5.8 ± 0.3	n.s.	5.1 ± 0.3	5.6 ± 0.3	n.s.
Inactive thymulin (pg/ml)	P	45.0 ± 9.6	0.6 ± 0.2	<0.001	33.5 ± 10.0	0.5 ± 0.1	<0.001
Total PBL (absolute no./mm^3)	B	2115 ± 205	3220 ± 210	<0.05	2100 ± 209	2917 ± 215	<0.05
$CD3^+$ (absolute no./mm^3)	B	1530 ± 325	2320 ± 210	<0.05	1335 ± 307	2225 ± 210	<0.05
$CD4^+$ (absolute no./mm^3)	B	n.d.	n.d.		508 ± 203	1107 ± 203	<0.05
$CD8^+$ (absolute no./mm^3)	B	n.d.	n.d.		610 ± 310	750 ± 213	n.s.
PHA (cpm × 10^{-3} cells)	L	70.5 ± 12.7	82.3 ± 15.5	n.s.	n.d.	n.d.	
ConA (cpm × 10^{-3} cells)	L	47.3 ± 11.4	46.5 ± 10.5	n.s.	n.d.	n.d.	

Mean ± SD; p calculated by paired Student's t-test. Zinc (1 mg Zn^{2+}/kg body weight per day) was administered through two cycles (two months each) separated by a period of 10 months. P = plasma; B = blood; L = lymphocytes; PBL = peripheral blood lymphocytes.

before and after nutritional intervention. A summary of the effects observed in a group of 25 DS subjects (age range = 6–15 years) should include the following (Table 5):

1. Zinc supplementation corrects the marginal zinc deficiency present in DS subjects without altering plasma copper concentrations (7); zinc plasma levels after zinc supplementation are even higher than those of age-matched healthy controls. A dramatic increase in plasma active thymulin is also recorded in DS children after zinc supplementation to values near those found in normal children. Concomitantly, inactive zinc-non-bound FTS molecules completely disappear from the plasma of DS subjects at the end of zinc supplementation (Table 5) . These findings clearly support the idea that the defect in thymic hormone activity is not due to thymic failure but to a low bioavailability of zinc ions in DS subjects (7, 36, 49).
2. With regard to peripheral immune efficiency, zinc supplementation in DS subjects increases the number of total peripheral T-lymphocytes ($CD3^+$) and T-helper/inducer cells ($CD4^+$), whereas the absolute number of T-suppressor/citotoxic cells ($CD8^+$) remains unaffected. Zinc supplementation also slightly increases lymphocyte responsiveness to PHA, without reaching statistical significance, however, while ConA responsiveness is unaffected (7).
3. With regard to hormonal profile, plasma thyrotropin levels decrease after zinc supplementation to normal values, while rT_3 plasma levels increase to the values recorded in healthy controls (Table 5) (22). Basal plasma growth hormone (GH) levels, on the contrary, do not show significant increments after zinc supplementation (68); more precise measurements, however, are required in order to assess GH turnover under zinc supplementation.
4. A marked decrease in the number of infectious episodes (mainly in the

Table 6. Effect of zinc supplementation on pituitary–thyroid function and on frequency of illness

	Before[a]	After[b]	p[c]
TSH (μU/ml)	6.80 ± 1.25	1.97 ± 1.15	<0.001
rT_3 (ng/ml)	0.08 ± 0.11	0.22 ± 0.12	<0.01
T_3 (ng/dl)	147.00 ± 8.30	135.00 ± 7.20	n.s.
T_4 (μg/dl)	18.00 ± 0.30	17.80 ± 0.40	n.s.
Number of infective episodes	2.90 ± 0.60	1.55 ± 0.20	⩽0.05
Days with elevated body temperature	9.30 ± 3.01	3.75 ± 1.18	⩽0.04
Days absent from school	10.80 ± 3.50	5.50 ± 1.90	⩽0.05

For this cycle (3rd) of zinc supplementation, zinc was given for four months.
[a] Clinical evaluation refers to the 12-month period preceding zinc supplementation.
[b] Clinical evaluation refers to the 12-month period following zinc supplementation.
[c] Statistical analysis was performed according to the two-tailed and paired Student's *t*-test.
Modified from ref. 22. Reproduced by permission of the Society for Neuroscience.

respiratory tract) and in the number of days with elevated body temperature was observed after zinc supplementation. Such a decrement is associated with an increased (50%) school attendance (Table 6) (22). The action of zinc supplementation on all the laboratory parameters taken into consideration is not long-lasting: DS subjects, after 8–10 months of interruption of zinc supplementation, resume the pretreatment values (Table 5). No side effects of zinc supplementation were recorded.

CONCLUSIONS AND PERSPECTIVES

On the basis of the data reported above it may be deduced that many of the immune and neuroendocrine alterations in DS may be ascribed to a common pathogenic event rather than to different causes.

The observation that zinc supplementation in DS subjects is able to restore thymic endocrine activity, various immune parameters and also some hormonal factors clearly pinpoints the relevance of zinc in the maintenance of various body homeostatic mechanisms.

If these findings clearly support the view that zinc supplementation is a valid approach to correct some immunological and neuroendocrine abnormalities in DS subjects, they should be considered still fragmentary since the widespread action of zinc (64–70) may influence many other physiological functions which have not yet been tested.

On these premises we plan to verify the role played by zinc supplementation in the complex psycho-neuroendocrine–immune network by setting up a clinical trial with an integrated analysis of the effect of zinc supplementation, not only in the neuroendocrine–immune interactions, but also in the other domains of the network and in particular in the psychoneurological domain.

In this context we plan to use neuropsychological batteries, particularly suitable for studying single or specific cognitive functions, like attention and memory. Neuropsychological batteries are well known in neuropsychology but they have not been extensively used to study subjects with mental retardation. The right choice of instruments suitable for studying cognitive functions in these conditions is required. In particular, for DS children during the first period of life it is important to study visual attention. Visual evoked potential and quantitative EEG analysis may further offer relevant instruments for neurophysiological analysis.

These neuropsychological batteries should be administered together with tests suitable for studying emotional aspects, because we know that many of these aspects may interact with cognitive functions. A knowledge of intellectual performance may be relevant also for the definition of biomarkers of aging in subjects with mental retardation.

This interdisciplinary approach may not only favor a better understanding of the role of zinc in DS subjects, but may also offer an integrated model for the comprehension of the syndrome and for the choice of educational, rehabilitation and therapeutic instruments.

REFERENCES

1. Lejeune J. Le mongolisme: premier example d'aberration autosomique humaine. *Ann Genet* 1959; **1:** 41–49.
2. Martin GM. Genetic syndrome in man with potential relevance to the pathology of aging. In *Genetic Effects on Aging*, edited by Bergsma D, Harrison DE, National Foundation March of Dimes, Birth Defects, Liss, London, 1978; **14:** pp. 5–39.
3. Baird PA, Sadovnick AD. Causes of death according to age in Down's syndrome. *Am J Hum Genet* 1988; **43:** 239–248.
4. Ugazio AG, Maccario R, Notarangelo LD, Burgio GR. Immunology of Down's syndrome: a review. *Am J Med Genet* (suppl.) 1990; **7:** 204–212.
5. Walford RY. Immunological studies of Down's syndrome and Alzheimer's disease. *Ann NY Acad Sci* 1982; **396:** 95–106.
6. Lockitch G, Sing VK, Puterman ML, Godolphin WJ, Sheps S, Tingle AJ, Wong F, Quigley G. Age-related changes in humoral and cell-mediated immunity in Down's syndrome children living at home. *Pediatr Res* 1987; **22:** 536–540.
7. Franceschi C, Chiricolo M, Licastro F, Zannotti M, Masi M, Mocchegiani E, Fabris N. Oral zinc supplementation in Down's syndrome: restoration of thymic endocrine activity and of some immune defects. *J Ment Defic Res* 1988; **32:** 169–181.
8. Cossarizza A, Monti D, Montagnani G, Ortomlani C, Masi M, Zannotti M, Franceschi C. Precocious aging of the immune system in Down's syndrome: alteration of B lymphocytes, T lymphocyte subsets and cells with natural killer markers. *Am J Med Genet* (suppl.) 1990; **7:** 213–218.
9. Bertotto A, Crupi S, Arcangeli C, Gerli R, Marinelli I, Velardi A, Vaccaro R. T cell response to anti CD2 monoclonal antibodies in Down's syndrome. *Scand J Immunol* 1989; **30:** 39–43.
10. Nair MPM, Schawartz SA. Association of decreased T cell mediated natural cytotoxicity and interferon production in Down's syndrome. *J Clin Immunol Immunopathol* 1984; **33:** 412–424.
11. Barking RM, Weston WL, Humber JR, Maire F. Phagocytic function in Down's syndrome. I. Chemotaxis. *J Ment Defic Res* 1980; **24:** 243–250.
12. Barking RM, Weston WL, Humber JR, Sunada K. Phagocytic function in Down's syndrome. II. Bactericidal activity and phagocytosis. *J Ment Defic Res* 1980; **24:** 251–256.
13. Griffin WS, Stanley LC, Ling C, White L, MacLeod V, Perrot LJ, White CL, Araoz C. Brain interleukin 1 and S-100, immunoreactivity are elevated in Down's syndrome and Alzheimer's disease. *Proc Natl Acad Sci USA* 1989; **86:** 7611–7615.
14. Taylor GM, Williams A, D'Sousa SW. Increased expression of lymphocyte functional antigen in Down's syndrome. *Lancet* 1986; **ii:** 740–742.
15. Burgio GR, Ugazio AG, Nespoli L, Maccario R. Derangements of immunoglobulin levels, phytohemaglutinin responsiveness and T and B cell markers in Down's syndrome at different ages. *Eur J Immunol* 1975; **5:** 600–603.
16. Burgio GR, Ugazio AG, Nespoli L, Maccario R. Down's syndrome: a model of immunodeficiency. *Birth Defects* 1983; **19:** 325–327.
17. Levin S. The immune system and susceptibility to infections in Down's syndrome.

In *Oncology and Immunology of Down's Syndrome*, edited by McCoy E, Epstein C, Liss, New York, 1987; pp. 143–162.
18. Avanzini MA, Soderstrom T, Whal M, Plebani A, Burgio GR, Hanson LA. IgG subclass deficiency in patients with Down's syndrome and aberrant hepatitis B vaccine response. *Scand J Immunol* 1988; **28:** 465–470.
19. Annerén G, Sara VR, Hall K, Tuvemo T. Growth and somatomedin response to growth hormone in children with Down's syndrome. *Arch Dis Child* 1986; **61:** 48–52.
20. Annerén G, Gustavson KH, Sara VR, Tuvemo T. Growth retardation in Down's syndrome in relation to insulin-like growth factors and growth hormone. *Am J Med Genet* (Suppl.) 1990; **7:** 59–62.
21. Lejeune J, Peeters M, Deblois MC, Bergeire M, Grillot A, Rethorte MU, Vallee G, Inzembart M, Devaux JP. Thyroid function and trisomy 21: TSH increase and rT3 deficiency. *Ann Genet* 1988; **31:** 137–143.
22. Licastro F, Mocchegiani E, Zannotti M, Arena G, Masi M, Fabris N. Zinc affects thyroid hormones in children with Down's syndrome: normalization of thyroid stimulating hormone and of reversal triiodothyronine plasmic levels by dietary zinc supplementation. *Int J Neurosci* 1992; **65:** 259–268.
23. Pueschel SM, Pezzullo JC. Thyroid dysfunction in Down's syndrome. *Am J Dis Child* 1985; **139:** 636–639.
24. Cutler AT, Benezra-Obeiter R, Brink SJ. Thyroid function in young children with Down's syndrome. *Am J Dis Child* 1986; **140:** 479–483.
25. Hestnes A, Stovner LJ, Husoy O, Folling I, Fougner KJ, Sjaastad O. Hormonal and biochemical disturbance in Down's syndrome. *J Ment Defic Res* 1991; **35:** 179–193.
26. Radetti G, Drei F, Betterle C, Mengarda G. Down's syndrome, hypothyroidism and insulin-dependent diabetes mellitus. *Helv Pediatr Acta* 1986; **41** 377–380.
27. Hsiang YG, Berkowitz GD, Bland GL, Migeon CJ, Warren AC. Gonadal function in patients with Down's syndrome. *Am J Med Genet* 1987; **27:** 449–458.
28. Kars H, Broekema W, Glaudemans I, Vehoeven WM, Vanree JM. Naltrexone attenuates self-injurious behaviour in mentally retarded children. *Biol Psychiatry* 1990; **27:** 741–746.
29. Robb SA, Vincent A, McGregor MA, McGregor AM, Newson-Davis JM. Achetylcholine receptor antibodies in the elderly and in Down's syndrome. *J Neuroimmunol* 1985; **9:** 139–146.
30. Mann DM, Yates PO, Marcyniuk B, Ravindra CR. Pathological evidence for neurotransmitter deficits in Down's syndrome of middle age. *J Ment Defic Res* 1985; **29:** 125–135.
31. Udeschini G, Casati G, Bassani F, Picotti CB, Culotta P. Plasma cathecolamines in Down's syndrome, at rest and during sympathetic stimulation. *J Neurol Neurosurg Psychiatry* 1985; **48:** 1060–1062.
32. Pierotti AR, Harmar AJ, Simpson J, Yates CM. High-molecular-weight forms of somatostatin are reduced in Alzheimer's disease and Down's syndrome. *Neurosci Lett* 1986; **63:** 141–146.
33. Godridge H, Reynolds GP, Czudek C, Calcutt NA, Benton M. Alzheimer-like transmitter deficits in adult Down's syndrome brain tissue. *J Neurol Neurosurg Psychiatry* 1987; **50:** 775–778.
34. Schapiro MB, Kay AD, May C, Ryner AK, Haxby JV, Kaufman S, Milstien S, Rapoport SI. Cerebrospinal monoamines in Down's syndrome adults at different ages. *J Ment Defic Res* 1987; **31:** 259–269.
35. Hamburger-Bar R, Eisenberg J, Belmaker RH. Animal and clinical studies of vasopressin effects on learning and memory. *Isr J Med Sci* 1987; **23:** 12–18.

36. Fabris N, Mocchegiani E, Amadio L, Zannotti M, Licastro F, Franceschi C. Thymic hormone deficiency in normal aging and Down's syndrome: is there a primary failure of the thymus? *Lancet* 1984; **i:** 983–986.
37. Annerén G, Johansson E, Lindh U. Trace element profile in individual blood cells from patients with Down's syndrome. *Acta Pediatr Scand* 1985; **74:** 259–263.
38. Purice M, Maximilian C, Dimitru I, Ioan D. Zinc and copper in plasma erythrocytes of Down's syndrome children. *Endocrinologie* 1988; **26:** 113–117.
39. Bruhl HH, Foni J, Lee YH, Madow A. Plasma concentrations of magnesium, lead, lithium, copper and zinc in mentally retarded persons. *Am J Ment Defic* 1987; **92:** 103–111.
40. Bjorkstein B, Back U, Gustavson H, Hallmans G, Haggiof B, Tarvnik A. Zinc and immune function in Down's syndrome. *Acta Pediatr Scand* 1980; **69:** 183–187.
41. Ganrot PO. Metabolism and possible health effects of aluminium. *Env Health Perspect* 1986; **65:** 363–441.
42. Farrar G, Altman P, Welch S, Wychrij O, Ghose B, Lejeune J, Cobert J, Prasher V, Blair JA. Defective gallium-transferrin binding in Alzheimer's disease and Down's syndrome: possible mechanism for accumulation of aluminium in brain. *Lancet* 1990; **335:** 747–750.
43. Nalini K, Aroor AR, Rao A. Urinary levels of taurine in mentally retarded children. *J Ment Defic Res* 1989; **33:** 271–274.
44. Coleman M, Sorbel S, Bhagavan HN, Coursin D, Marquardt A, Guay M, Hunt C. Double binding of Vit B6 in Down's syndrome infants. Part 1. Clinical and biochemical results. *J Ment Defic Res* 1985; **29:** 233–240.
45. Cartlidge PH, Curnock DA. Specific malabsorption of Vit B12 in Down's syndrome. *Arch Dis Child* 1986; **61:** 514–515.
46. Metcalfe T, Bowen DM, Muller DP. Vitamin E concentrations in human brain of patients with Alzheimer's disease, fetuses with Down's syndrome, centenarians and controls. *Neurochem Res* 1989; **14:** 1209–1214.
47. Colombo ML, Girardo E, Incarbone E, Conti R, Ricci BM, Maina D. Vitamin C in children with trisomy 21. *Minerva Pediatr* 1989; **41:** 189–192.
48. Fabris N, Mocchegiani E, Muzzioli M, Provinciali M. Neuroendocrine thymus interactions: perspectives for intervention in aging. *Ann NY Acad Sci* 1988; **521:** 72–87.
49. Fabris N, Mocchegiani E, Muzzioli M, Provinciali M. The role of zinc in neuroendocrine–immune interactions during aging. *Ann NY Acad Sci* 1991; **621:** 314–326.
50. Deaton JG. The mortality rate and causes of death among institutionalized mongols in Texas. *J Ment Defic Res* 1973; **17:** 117–123.
51. Knox GE, Bensel RW. Gastrointestinal malformations in Down's syndrome. *Minn Med* 1972; **55:** 542–544.
52. Pueschel SM. Clinical aspects of Down's syndrome from infancy to adulthood. *Am J Med Genet* (Suppl.) 1990; **7:** 56–60.
53. Sutnik AI, London T, Blumber BS, Gerstley BJS. Susceptibility to leukemia: immunologic factors in Down's syndrome *J Natal Cancer Inst* 1971; **47:** 923–933.
54. Bertotto A, Crupi S, Arcangeli C, Gerli R, Scalise F, Fabietti G, Agea E, Vaccaro R. T-cell response to phorbol ester PMA and calcium ionophore A23187 in Down's syndrome. *Scand J Immunol* 1989; **30:** 583–586.
55. Ugazio AG, Lanzavecchia A, Jayakar S, Plebani A, Duse M, Burgio GR. Immunodeficiency in Down's syndrome. *Acta Pediatr Scand* 1978; **67:** 705–708.
56. Gordon MC, Sinha SK, Carlson SD. Antibody response to influenza vaccine in patients with Down's syndrome. *Am J Ment Defic* 1971; **75:** 391–399.

57. Levin S, Schlesinger M, Handzel Z, Hahm T, Altman Y, Czenobiski B, Boss Y. Thymic deficiency in Down's syndrome. *Pediatrics* 1979; **63:** 80–83.
58. Musiani P, Valitutti S, Castellino F, Larocca LM, Maggiano N, Piantelli M. Intrathymic deficient expansion of T cell precursors in Down's syndrome. *Am J Med Genet* (Suppl.) 1990; **7:** 219–224.
59. Murphy M, Lempert MJ, Epstein LB . Decreased level of T cell receptor expansion by Down's syndrome (trisomy 21) thymocytes. *Am J Med Genet* (Suppl.) 1990; **7:** 234–237.
60. Dardenne M, Pleau JM, Nabama B, Lefancier M, Denien J, Choay J, Bach JF. Contribution of zinc and other metals to the biological activity of the serum thymic factor. *Proc Natl Acad Sci USA* 1982; **79:** 5370–5373.
61. Torrado C, Bastian W, Wisnieski KE, Castells S. Treatment of children with Down's syndrome and growth retardation with recombinant human growth hormone. *J Pediatr* 1991; **119:** 478–483.
62. Wisnieski KE, Laure-Kamionowska M, Connel F, Yen GY. Neuronal density and synaptogenesis in the postnatal stage of brain maturation in Down's syndrome. In *The Neurobiology of Down's Syndrome*, edited by Epstein CJ, Raven Press, New York, 1986; pp. 29–44.
63. Wisnieski KE. Down's syndrome children often have brain with maturation delay, retardation growth and cortical dysgenesis. *Am J Med Genet* (Suppl.) 1990; **7:** 274–281.
64. Pfeiffer CC, Braveman ER. Zinc, the brain and behaviour. *Biol Psychiatry* 1982; **17:** 513–532.
65. Prasad AS. Clinical, endocrinological and biochemical effects of zinc deficiency. *Clin Endocrinol Metab* 1985; **14:** 567–589.
66. Dunn MF, Pattison SE, Storm MC, Quiel E. Comparision of the zinc binding domains in the 7S nerve growth factor and the zinc-insulin examer. *Biochemistry* 1980; **19:** 718–725.
67. Scuderi P. Differential effect of copper and zinc on human peripheral macrocyte cytokine secretion. *Cell Immunol* 1990; **126:** 391–405.
68. Napolitano G, Palka G, Grimaldi S, Giuliani C, Laglia G, Calabrese G, Satta MA, Neri G. Growth delay in Down's syndrome and zinc sulphate supplementation. *Am J Med Genet* (Suppl.) 1990; **7:** 63–65.
69. Editorial. Essential trace element and thyroid hormones. *Lancet* 1992; **339:** 1575–1576.
70. Irving LP. Zinc deficiency and visual impairment. *Am J Ophthalmol* 1978; **85:** 871–875.

Discussion of Presentation

GIORGIO ALBERTINI
(Edited by D.A. Snider and Salvador Castells)

G. Annerén: We have studied zinc, copper and selenium together with acute-phase proteins. It is well known that if you have an infectious disease, the zinc drops and the copper is increasing. This is what we see in Down's syndrome and there is a good correlation with acute-phase proteins. Do you look at acute-phase proteins?

G. Albertini: A careful analysis is in progress. α_2-Macroglobulin binds zinc with high K_d, and therefore its increase may interfere with the bioavailability of the element for molecules, such as thymulin, with lower K_d.

G. Annerén: In Down's syndrome, about 50% of the children are prone to infectious diseases. The number of infections drops over time. Do you have a control group which you follow at the same time, without treatment for the first period and the second period?

G. Albertini: No, we did not have a control Down's syndrome group, but only a control group of healthy age-matched children (school companions) who showed the same frequency of infectious diseases in the two-year period considered and which was similar to that of Down's syndrome subjects after zinc supplementation.

A. Pfadt: We did a literature search recently on factors that are associated with severe pica behavior in mentally retarded individuals and we found zinc deficiency to be commonly reported, but we find resistance by professionals and caregivers to give zinc supplementation therapy. Is there any basis for concern about treating pica with zinc supplementation?

G. Albertini: We give zinc supplementation only to those who are zinc deficient based on the analysis of thymulin zinc saturation, but most of the Down's syndrome children (>90%) show altered zinc turnover. No side effects have been recorded. It has to be underlined that in the case of Down's syndrome children zinc supplementation should not be considered as an integration of the diet, but as a correction of a defect. Obviously, the impact of zinc on the complex nervous–neuroendocrine–immune network still requires further study. We think we are at the beginning of a very important link between three homeostatic systems. We have published three or four studies, but we have to improve the methods to determine the role of zinc in the homeostatic systems: central nervous, neuroendocrine and immune systems.

T. Sharav: You said you were studying thyroid TSH levels in children 2–14 years of age. I found in my studies that with or without zinc TSH levels tend to decrease with age. I think this may also be true of rT_3. I'm wondering if this is a zinc effect or a purely physiological effect. If you would have gone into the pubertal group you might have started getting high TSH levels because of this autoimmune type of hypothyroidism, so you need to look at age.

G. Albertini: The correlation between zinc levels and thyroid and TSH levels is very high, and we don't think the relationship is simply due to age. In all cases the TSH–rT_3 alteration has been observed in Down's syndrome subjects, 6–16 years of age.

Z. Laron: The question we come back to is: whether the zinc effect is a specific effect or a transitory age-related effect. A more controlled study is needed to show one or the other.

N. Stahnke: You show rT_3. What is the clinical significance of low rT_3 levels prior to treatment, and higher levels during treatment, for this hormone is biologically inactive? Thus, the higher level of an inactive hormone does't seem to be of any benefit to the patient.

G. Albertini: Whether rT_3 has any biological role is still controversial. According to some authors, rT_3 may represent a feedback mechanism of thyroid hormone. These findings have been obtained, however, in experimental animals. In humans and particularly in Down's syndrome subjects, the role of rT_3 remains to be established.

17 Summary Discussion of Presentations of the First Day

ZVI LARON, (moderator)
Salvador Castells and D.A. Snider (editors)

Z. Laron: I would like to divide the discussion into several parts. First, what is the effect of GRP, growth hormone-releasing hormone (GHRH), GH and insulin-like growth factor I (IGF-1) on central nervous system (CNS) development and maturation. There has been evidence presented here that GH, IGF-1 and GRP affect *in utero* brain development or postnatal maturation. Are you convinced by what has been presented?

N. Stahnke: If we believe in a strong relationship between GH secretion and brain development, we should expect patients with congenital, idiopathic human growth hormone (hGH) deficiency to be mentally retarded, but this is obviously not the case. Thus, doubt arises concerning the relationship between these two factors.

V. Sara: There is a great doubt in my mind that GH can have any direct influence on intrauterine brain development. However, there is absolutely no doubt in my mind that IGF-1 has a role in brain development. I think in fact that it remains to be demonstrated that GH, by itself, can directly influence intrauterine growth development. *In vitro*, GH has absolutely no effect on the proliferation of neural or glial cells or on the differentiation of those cells. If you also go to the transgenic GH animals, you do not see any change in growth and you see absolutely no change in the expression of IGF-1 genes during embryonic life. GH only starts to regulate IGF-1 expression at about three weeks of postnatal age, which would be about three to six months in the humans when it starts to play a role in IGF synthesis. Prior to that synthesis of IGF is not regulated by GH.

Z. Laron: In other words, *in vitro*, GH has no effect?

V. Sara: Those cells *in vitro* are producing IGF-1 and IGF-2; it is simply that production cannot be regulated by GH at that stage of their development. To my knowledge, only one population of cells in the brain shows a receptor for GH, and one of these cells populations is a very discrete population of Purkinje cells in the vermis of the cerebellum. This is the only really strong evidence for these cells. There is other evidence which really has to be documented by other groups, and that is that there is a GH receptor in the hippocampus and the hypothalamic region. If these are the only cell populations showing receptors for the hormone, what is the hormone doing in the

Growth Hormone Treatment in Down's Syndrome. Edited by S. Castells and K.E. Wisniewski

brain? What is of interest is that these Purkinje cells in the cerebellum may be the same population expressing IGF-1. This is really the only situation I am aware of where there may be a role for GH in the brain; and this is the adult cerebellum. My question is: in Laron-type dwarfism (LTD) do you see any type of motor disturbance at all which could be explained by this one cell population?

Z. Laron: They had a very slow motor development.

G. Merriam: Two comments, and perhaps they point in different directions. One, peripheral levels of IGF-1 may be the least relevant compartment to correlate with CNS effects in development or in adulthood. Two, I think it is possible to overstate the role of GH as a regulator of IGF_2-1 production. Several groups have shown that nutritional adequacy is a much stronger modulator of IGF production. In normal subjects, all you have to do is fast them for 24 h, and you can turn off hepatic production of IGF-1 while GH is perfectly sufficient. Thus some GH is a permissive factor for IGF-1. Many other factors regulate IGF-1 and those factors differ from tissue to tissue. I think we should be very cautious about generalizing peripheral effects to CNS effects.

Z. Laron: If we agree that IGF-1 has an effect, and we implicate that a deficiency of IGF-1 has primarily a negative effect, what you mentioned about the GHRH may have a regulatory effect in LTD patients; they have very low IGF-1, so GHRH and GH are very high. Now is there any doubt that overstimulation or pharmacological application of one of the peptides you have mentioned will affect sleep or other neurological functions?

G. Merriam: I don't know. Since it is not ethical to overstimulate directly, you have to look at experiments of nature, such as children with gigantism, including patients who had ectopic tumors producing ectopic GHRH in enormous quantities accelerated growth and gigantism with sky-high levels of IGF-1 peripherally. I don't know if the Charlottesville patient or the one from Lyon, which were characterized early for that, had neurological studies and performed, but cases like this would be a place to test these ideas.

Z. Laron: What about the transgenic mice that had tremendous amounts of GHRH?

M.L. Oster-Granite: The only thing that has been documented so far about GH transgenics is in metallothionine transgenics—the gene is not over expressed until you drive the system postnatally by feeding metals. In those animals, you can show you can create a nervous system that has more neurons. The question you have to ask yourself is whether there is an allometric correlation between the size of the brain and the size of the body in those animals. To my knowledge, the correlation is the mirror image of what you would see in a dwarfed individual; where there are about 65% of the normal number of neurons and the body is only 35% as large. The giant mice have about a 50% larger body, but they may have only about 35%

more neurons. The question that has to be asked, at this point in time, is whether they have more neurons because they really make more in the first place, or whether it is because the process of natural cell death is altered. Behaviorally, I don't think a lot has been done with these transgenic mice; I can assure you that they are not smarter. They may age prematurely.

Z. Laron: Do they have more IGF-1?

M.L. Oster-Granite: I have no idea.

V. Sara: They have more IGF-1 after about three weeks of postnatal life.

Z. Laron: I am asking if GHRH secretion in the brain causes more IGF-1 in the brains in these animals.

V. Sara: No one has looked. IGF-1 transgenics show an abnormal myelination from the overexpression of IGF-1 in the brain.

Z. Laron: It seems from the discussion that IGF-1 has been shown to have a role in brain development and in the cells. What can we say about the binding proteins in the brain? There is certainly an effect of IGF on the binding proteins. Is the IGF-1 effect modulated by local IGF binding proteins (IGFBPs)? I don't think there is IGFBP-3 in the brain. The only studies that have been done are in the rat, and there is clear evidence that IGFBP-2 and IGFBP-6 are being produced; they seem to be the two major binding proteins in the CSF. They are being produced during early development in the rat nervous system. From *in situ* hybridization they seem to localize in the cells, which are adjacent to the cells that are producing IGF-1 it seems they are being secreted picking up the IGF, and there is *in vitro* evidence that this is a modulating biological activity; furthermore, the IGF activity is not only a paracrine but also an endocrine activity.

V. Sara: Major binding proteins in the brain are IGFBP-2 and IGFBP-6, which bind IGF-2 better than IGF-1. This means they have a selective affinity for IGF-2; they pick it up and it circulates in the CSF. You don't find IGF-1 in the CSF, and the reason is that I think IGF-1 is having a very local autocrine or paracrine action, whereas I think IGF-2 is being picked up by binding proteins more and carried through the CSF, which is the CNS. By this I mean IGF-2 is delivered to other sites in the CNS.

Z. Laron: IGFBP-2 and IGF-2 are very high in the CNS. If GH has no action on the brain and IGFBP-3 is largely GH dependent, how would you imply the presence of IGFBP-3?

V. Sara: Some researchers think it is present, but I think we need more evidence concerning IGFBP-6 and IGFBP-2, which are the major CSF-binding proteins and are certainly not regulated by GH. Just to add that IGF-2 and IGFBP-2 and IGFBP-6 are made predominantly in the choroid plexus, which would give them access to the circulation of the CNS.

Z. Laron: There is evidence that IGFBP-3 may also be regulated by IGF-1 [Kaneti *et al.*, *Acta Endocrinol.*, 1993; **28:** 144–149] and not only GH. I wish to ask: are some Down's syndrome patients GH deficient? And are some Down's syndrome patients hypothyroid at a certain age?

G. Merriam: If they are hypothyroid, then you cannot successfully do GH issue of whether their body mass index (BMI) has been factored in is relevant. I was impressed looking at Dr Castells' presentations of GH levels in provocative tests that were done by his group. A relatively large percentage had very high basal GH levels, even before the stimulus was applied in some cases —20 ng/ml or more. GH is a stress hormone, and if you have just stimulated a pulse of growth release there is a relative refractoriness to getting a second pulse. I wonder whether, in a group of children with mental retardation, the medical process might be stressful or whether the circumstances in which the GH testing is done would allow one to know that baseline conditions have been reached before doing the testing. If not, one might not be able to elevate GH further. The very modest decrement in IGF-1 levels that was shown suggests that, if there is a GH deficiency in the aggregate as a population, it may not amount to too much.

The question for therapies is, does the response to GH testing really matter if you are considering the various potential benefits of GH therapy? The large database that has been accumulated in the Genentech collaborative trial has shown very little correlation between the magnitude of the response to GH provocative testing, to GH testing, and the response to administered GH in doses similar to those described here. So while there may be physiological interest into insights into the neuropathology of Down's syndrome, I'm not sure that we really need to know that answer in growth-retarded Down's syndrome children to consider them as candidates for GH therapy.

Z. Laron: To boil this down, Dr Merriam has said that many people taking the tests are stressed; he would have liked to have had a pretest rest period in order to make sure that these people were less stressed.

G. Merriam: For example, I am referred a lot of patients with hyperprolactinemia and reproductive disorders. In these patients, if we place a heparin lock and let them sit quietly for 90 min and then initiate the testing, prolactin basal concentrations are found to return to within normal limits in about 30–40% of the patients studied. I really think that a period of resting after the initial stressful venipuncture might be useful to avoid exhausting the GH reserve, before you beat the stimulation test.

Z. Laron: None of the patients has complete GH deficiency, and the variation in so-called low values may stem from the difference in antisera used, which we know may vary. Would you consider that serum IGF-1 levels would be a better measure to see if the Down's syndrome patients are GH deficient? The exception has to be very young children in whom IGF is low anyway.

G. Merriam: I'm not certain. In a situation where normal values change, and it is unclear whether we should consider chronological age, bone age, or height age, and where the abnormality is slight, I am not sure 1GF-1 is a better probe. I think that careful assessment of GH secretion in the way that various groups here have suggested doing is the way to go. In other common chromosomal disorders associated with growth retardation, such as Turner's syndrome, GH deficiency is not complete. In Turner's syndrome, there is some evidence of GH deficiency , but the trials of GH therapy, which are proceeding in Turner's syndrome, went ahead largely realizing that it didn't matter if they were GH deficient or not.

Z. Laron: I would like to pinpoint the question whether Down's syndrome patients are GH deficient.

S. Castells: As you are aware that several parameters for GH deficiency were evaluated by us. One involved specific tests for GH deficiency. We did 24 and 12 h integrated GH concentrations and clonidine and L-dopa stimulation tests, which are tests of hypothalamic function; and we did pituitary testing with GHRH. Another biochemical parameter was IGF-1 concentration before treatment, and every three months during rhGH treatment. We also assessed bone age, nutrition, growth rate per year and sexual maturation. When we look at the annual growth rate, we see that these children have not been growing well from the age of a few months up until puberty—a pattern of growth that fits partial GH deficiency. When we look at bone age and see retarded bone age with normal thyroid function, we have another parameter of partial GH deficiency. If you put together the results of the biochemical tests, the growth retardation, the growth velocity per year, and the retarded bone age together, you have many parameters indicating partial GH deficiency. Finally, acceleration of growth velocity with rhGH treatment for five years, as presented by us, is another finding suggestive of GH deficiency. Thus we used in these children classical clinical and biochemical parameters to diagnose GH deficiency.

Z. Laron: Bone age is a very non-specific parameter.

S. Castells: I agree, but it was another parameter taken into account to make a decision to treat with GH. We not only look at the testing itself, but we took into consideration clinical, radiological and biochemical parameters that go with GH deficiency. The second point some people feel, as with Turner's syndrome, is what does it matter what the testing shows? If growth velocity accelerates with rhGH treatment, they must have insufficient GH production. It has been suggested that a trial test of GH for a period of six months or one year may be important to determine if a deficiency of GH is present. Our data on rhGH treatment showed very clearly that GH is effective in treating short stature in Down's syndrome.

Z. Laron: Would you agree, Dr Castells, that your data do not show that Down's syndrome patients have a clear GH deficiency? If I understood your data, they have a dysregulation—some kind of a CNS abnormality—which may be like neuroregulatory dysfunction, but they don't have true GH deficiency; there were no patients who had a repeated level of less than 5 ng/ml. This is very important, because in the next part of the discussion I would like to ask: in treating short children with Down's syndrome, are we treating GH deficient patients, or are we treating non-conventional, short children?

I think it would be very important to define these two categories. I am not denying that short children are being treated with GH and that some of them respond to therapy, even though they are not GH deficient, but we should decide to which category the Down's syndrome children belong.

S. Castells: Using the 24 h, 12 h testing, the four children that we tested showed a lack of significant peaks of GH, so it appears in this particular testing to be suggestive of low GH but not complete GH deficiency. There are some peaks there, but certainly smaller in number and amplitude than in the normal child. On the clonidine and L-dopa, out of 12 tested we had five patients with responses less than 10 ng/ml. Here there is disagreement between investigators on what is a normal response. The Canadians feel that 7 ng/ml is the right response, so that anyone that does not produce 7 ng/ml is abnormal, and anybody above 7 ng/ml is normal. In the USA, we believe that 10 ng/ml is a significant cut-off point between normals and abnormal. Now you suggest that 5 ng/ml may be the point between normal and abnormal response. So there is a significant disagreement in defining an abnormal response. But in our study, five of our patients did not have a normal response to both dopamine and clonidine. We feel that these patients should be considered GH deficient. We are left with seven patients who had a divergent response between clonidine and dopanine. In this case, we have a much weaker point in saying that they are GH deficient. We postulate that there is a difference between neuroendocrine dysfunction and hypopituitarism as each organ responds to stimuli. In the pituitary gland there is only one specific cell in the somatrope that can be stimulated to produce and release GH. When you are talking about the hypothalamus, you are talking about different cells responding to different neurotransmitters. The hypothalamus has specific cells to respond to dopamine, clonidine or hypoglycemia. Damage to one set of cells might not affect others, resulting in a dichotomy between dopamine and clonidine response.

G. Merriam: I would again like to caution that what you get in a study depends on your selection criteria. If one gets too hung up on the issue of GH deficiency and if you select patients to receive growth hormone based on GH testing, you may actually change the outcome. A more useful question to ask is: is GH supplementation of value in short children with Down's syndrome? While you are welcome to collect the GH provocative secretory or spontaneous data, from what I have heard of our knowledge base; I would not base the decision on giving growth hormone or not on the details of whether GH was 5 or 7 or 10 ng/ml, I think you'd get more useful long-term data just by doing a treatment study of short children with Down's syndrome.

A. Pfadt: Several investigators here have said that the assessment of GH deficiency is complicated by obesity in Down's syndrome, and I wanted to call people's attention to the literature on Prader-Willi syndrome, where GH secretory problems may be a contributor to obesity and there is evidence to suggest that GH supplementation or treatment can dramatically reduce obesity. I would suggest that the influence of GH can go both ways.

N. Stahnke: In my opinion, the point you raise, Dr Laron, is quite clear: from a clinical point of view, these patients do not have the classical complete isolated GH deficiency except in some rare cases. The clinical, auxological data are really not consistent with GH deficiency; this is in good line with the data presented by Dr Annerén, showing normal spontaneous GH secretion in these patients. That a short patient is growing better under GH therapy doesn't say anything about his GH secretion. We know from many studies, worldwide, that short patients grow better if you give GH. Think of the young acromegalics who grew very fast; they didn't have any GH deficiency, thus we should really keep in mind that patients with Down's syndrome do not have classical GH deficiency—that does not say anything about possible therapy with GH.

Z. Laron: I think there is more or less agreement from the data that exist today and the number of children tested that Down's syndrome patients do not have severe GH deficiency, or any deficiency at all. I would like to discuss one last point: do we agree that there are indications that GH treatment of children with Down's syndrome is promising to be beneficial and that such treatment should fall into the category of non-conventional GH therapy? Perhaps, as Dr Merriam suggests, we may treat short children with Down's syndrome. Dr Annerén says they are now treating small children at an early age, taking into consideration the possible affect on brain maturation GH may have via IGF, and are not taking into consideration whether the Down's syndrome patients are short. Is the dose adequate? We know that in many cases the dose has to be greater than it is in replacement therapy. Considering the above open questions, it seems to me that only well-controlled clinical trials are indicated at present, as Dr Annerén has mentioned.

18 Possible Complications of Growth Hormone Therapy in Down's Syndrome

SALVADOR CASTELLS[1]* AND D.A. SNIDER[2]
[1]Department of Pediatrics, State University of New York Health Science Center at Brooklyn, and [2]Institute for Basic Research in Developmental Disabilities, New York, USA

The conference conveners, Salvador Castells and Krystyna Wisniewski, have been most concerned about possible complications of growth hormone (GH) treatments in Down's syndrome (DS). They have built in safeguards concerning leukemia and hyperglycemia into their own protocol. They have followed blood sugars and glycohemoglobins every three months in these children because of the high frequency of diabetes in DS. In short, none of the 20 DS who have been receiving GH in their study have shown hyperglycemia or any increase in plasma glycohemoglobins. Thus they have some direct data available to rule out carbohydrate intolerance in DS under treatment with recombinant-human GH (rhGH) for up to five years at 1 ng/kg three days a week. The Castells and Wisniewski protocol is also sensitive to the possibility of leukemia in children with DS. Complete blood counts (CBCs) are taken every three months, and Castells and Wisniewski have not observed any changes in the CBC levels up to the five years of treatment with rhGH. While 20 children do not represent a large sample, given the fact that they have now treated some of the children for more than five years, it does have some value. Obviously, investigators must continue to monitor the children treated for possible complications.

The remainder of this chapter is taken from the discussion of complications during the conference. Those participating included Salvador Castells (Brooklyn, NY), Patrick Wilton (Stockholm, Sweden), Zvi Laron (Tel Aviv, Israel), and Gorän Annerén (Uppsala, Sweden). References relating to Castells' remarks have been added.

S. Castells: Immediately prior to the conference, Dr D.A. Snider received a letter from Dr Jo Anne Brasel, President of the Lawson Wilkins Pediatric

*Address for correspondence: Department of Pediatrics, SUNY Health Sciences Center of Brooklyn, 450 Clarkson Avenue, Brooklyn, NY 11203, USA.

Growth Hormone Treatment in Down's Syndrome. Edited by S. Castells and K.E. Wisniewski

Endocrine Society (LWPES), asking that the following points be made in the discussion: (a) that leukemia is more common in DS patients; (b) that leukemia has been reported in children from many countries, including the USA, who have been treated with GH for their hormone deficiency; and (c) that, therefore, there are pediatric endocrinologists who are concerned about whether such therapy might increase the leukemia risk in these patients. Further close follow-up over several years will be necessary to determine the extent, if any, of such risk.

The concern in the LWPES letter that leukemia is more common in DS patients emanates from epidemiological studies in the 1960s (1–4) that showed that persons with DS have a high incidence of leukemia. These studies were of institutionalized children and adults. Since most children with DS in the 1990s are now living with their families, the high incidence of leukemia found among institutionalized persons from the 1960s might not be found today given the improvements in standards of living experienced by children with DS today. If, for example, some types of leukemias can be induced by viruses, institutionalized DS individuals that had more viral diseases, like hepatitis, might have a higher incidence of leukemia than children with DS living at home. At the very least, the applicability of the studies in the 1960s to the present population of children with DS living at home is not clear. New epidemiological studies are needed of children with DS living at home, including those before the age of puberty, before the incidence of leukemia in this population can be evaluated.

The LWPES letter also raised a concern about the increased incidence of leukemia in isolated GH-deficient children receiving GH treatment. (A bibliography of studies on this issue follows this paper (5–25).) In 1988, the LWPES reviewed the data available (see reference 5, report of the International Workshop on Growth Hormone and Leukemia, 5 May 1988) and obtained an incidence of one case of leukemia for each 24 000 patient-years of risk. The expected incidence of leukemia in children under 15 years of age approximates to one per 42 000 of the population per year. The authors of that study concluded that: (a) there is a possible increase of leukemia in GH-deficient patients treated with GH; (b) we cannot conclude on the basis of current evidence that GH therapy is responsible for this possible increase; (c) from current estimates, patients can be told that if there is any increase in risk to an individual patient, it is small; and (d) all patients receiving GH should have close follow-up during and after therapy. It is our understanding that the conclusions and recommendations of the LWPES have not changed.

P. Wilton: As a pediatrician, it is important when you start to treat children that you don't do any harm. This is also important to pharmaceutical companies. We started with the traditional treatment of GH deficiency, and then we moved into the new field of treating young girls with Turner's syndrome. That is a common chromosomal aberration, and they certainly do not have classical GH deficiency. We know now that they grow. We were interested to see what could be the potential problems for this new group of non-GH-deficient persons. We have to think of what the Turner's syndrome girls have as problems: carbohydrate intolerance, autoimmune disorders, possibly malignant tumors; these were some of the problems on which we concentrated on studying during treatment. The carbohydrate problems have not shown up in the first eight years of GH treatment. We have seen no problems of autoimmunity in this group either. There have been no reports of leukemia in the Turner's syndrome girls. Now we are entering a new group: patients with DS. I agree with Dr Stahnke that they are not classical GH deficient. We have heard two presentations today about efficacy, including one of five years by Dr Castells. What could be the harm we are doing to such children? We have to look for potential problems: autoimmune problems, osteoporosis, carbohydrate intolerance (like diabetes mellitus) and leukemia. We have to concentrate our efforts here to look for safety in this group. Would you want to summarize the leukemia data in GH-deficient children, Dr Laron [Chairman of the session]?

Z. Laron: Being negative data at present, there seems to be no reason to enlarge on the issue. But since the treatment of DS syndrome patients is a non-conventional type of hGH treatment, maybe you could say a word about your experience with other types of growth retardation, and especially whether one should expect advancement of bone age with longer therapy?

P. Wilton: When it comes to dose, it seems to work very well; it is the common dose used for GH deficiency, so that when we treat girls with Turner's syndrome we usually treat with a higher dose than we have seen today. What we could look at is the outcome of the IGF. The values shown are in the range where you find fast-growing children. They aren't of great concern to me. When it comes to the question of whether GH accelerates the bone age, there has been a suggestion that this might occur. But the best way to prove or disprove this is to perform randomized clinical trials with one group non-treated, but carefully followed.

G. Annerén: I don't think it is time yet to recommend GH treatment except for controlled studies. We don't know what is happening with the skeletal age, and thus we don't know what the final height will be. If we find that a short period of GH treatment could restore normal growth velocity for life, long after rhGH as stopped, then I would say yes. If we could show that the mental development would be better, of course we recommend treatment. We do not have this information yet.

REFERENCES

1. Holland WW, Doll R, Carter CO. The mortality from leukemia and other cancers among patients with Down's syndrome (mongols) and among their parents. *Br J Cancer* 1962; **16:** 177–186

2. Jackson EW, Turner JH, Klauber MR, Morris FD. Down's syndrome: variation of leukemia occurrence in institutionalized populations. *J Chron Dis* 1968; **21:** 247–283.
3. Miller RW. Persons at exceptionally high risk of leukemia. *Cancer Res* 1967; **270:** 2420–2423.
4. Wald N. Leukemia associated with mongolism. *Lancet* 1961; **i:** 1288. 1961.
5. *Lawson Wilkins Pediatric Endocrine Society Meeting Minutes*, 5 May 1988.
6. Watanabe S *et al.* Leukemia in patients treated with growth hormone. *Lancet* 1988; **i:** 1159.
7. Delemarre-Van de Waal HA *et al.* Leukemia in patients treated with growth hormone. *Lancet* 1988; **i:** 1159.
8. Fisher DA *et al.* Leukemia in patients treated with growth hormone. *Lancet* 1988; **i:** 1159–1160.
9. Endo M, Kaneko Y, Shikano T *et al.* Possible association of human growth hormone treatment with an occurrence of acute myeloblastic leukemia with an inversion of chromosome 3 in a child of pituitary dwarfism. *Med Pediatr Oncol* 1988; **16:** 45–47.
10. Frisch H *et al.* Leukemia and growth hormone. *Lancet* 1988; **i:** 1335.
11. Ogilvy-Stuart AL, Ryder WD, Gattamaneni HR, Clayton PE, Shalet SM. Growth hormone and tumour recurrence. *Br Med J* 1992; **304:** 1601–1605.
12. Redman GP *et al.* Leukemia and growth hormone. *Lancet* 1988; **1:** 1335.
13. Sasaki U *et al.* Occurrence of acute lymphoblastic leukemia in a boy treated with growth hormone for growth retardation after irradiation of the brain tumor. *Jpn J Clin Oncol* 1988; **18:** 81–84.
14. Ogawa M *et al.* The occurrence of acute lymphoblastic leukemia shortly after the cessation of human growth hormone therapy. *Jpn J Clin Oncol* 1988; **18:** 255–260.
15. Hara T *et al.* Acute lymphoblastic leukemia in a patient with pituitary dwarfism under treatment with growth hormone. *Acta Pediatr Jpn* 1989; **31:** 73–77.
16. Santorio A *et al.* Previous human GH treatment and leukemia. *J Endocrinol Invest* 1989; **12** 131–132.
17. Stahnke N *et al.* Growth hormone therapy and leukemia. *Eur J Pediatr* 1989; **148:** 591–596.
18. Canadian Growth Hormone Advisory Committee. Growth hormone treatment and leukemia. *Canada Diseases Weekly Report*, 1989.
19. Ishii A *et al.* Chronic myelocytic leukemia probably promoted by growth hormone. *Tohoku J Exp Med* 1989; **158:** 263–264.
20. Wada E *et al.* Acute lymphoblastic leukemia following treatment with human growth hormone in a boy with possible preanemic Fanconi's anemia. *Jpn J Clin Oncol* 1989; **19:** 36–39.
21. Watanabe S *et al.* Risk factors for leukemia occurrence among growth hormone users. *Jpn J Cancer Res* 1989; **80:** 822–825.
22. Standen GR *et al.* Myelodysplastic syndrome with trisomy 8 in an adolescent with Fanconi anemia and selective IgA deficiency. *Am J Hematol* 1989; **31:** 280–283.
23. Boose ARM *et al.* Growth hormone therapy and leukemia. *Tijd Kindergeneesk* 1992; **60:** 1–6.
24. Blatt J *et al.* Leukemia in a child with a history of medulloblastoma. *Pediatr Hematol Oncol* 1991; **8:** 77–82.
25. Witherspoon RP *et al.* Secondary cancers after bone marrow transplantation for leukemia or aplastic anemia. *N Engl J Med* 1989; **321:** 784–789.

Part IV

ETHICAL, NURSING AND FAMILY ISSUES

19 Ethical Issues in the Use of Human Growth Hormone Treatments in Down's Syndrome

PHILIP R. REILLY
Shriver Center for Mental Retardation, Waltham, Massachusetts, USA

INTRODUCTION

Ethical analysis of any issue in the biomedical sciences requires that the analyst acknowledge the fundamental principles from which he proceeds. Since ethics is in essence the business of assessing facts in the context of values in order to decide what one ought to do, outcomes may and do vary tremendously among analysts depending on the principles they espouse. For one relevant example, recent cross-cultural studies of ethical issues in genetic counselling among clinicians in 18 nations discovered sharply differing views on the use of prenatal diagnosis and abortion as a method of sex selection (1). One can only make sense out of an assertion about what *ought* to be done if one knows the principles that the analyst brings to the decision-making process.

This modest ethical analysis does not proceed from a particular religious tradition or canon of law. Four universal principles shape the effort. Foremost is *autonomy*: the principle that each individual is of intrinsic worth and deserves moral respect simply by virtue of his or her existence. Next is the expectation that no person should knowingly act or fail to act in a manner that harms another individual. This is often called the principle of *non-maleficence*. Third is the understanding that actions or decisions not to act on behalf of another person should be beneficent, that is intended to do good. The principle of *beneficence* is the corollary of non-maleficence. The fourth principle is *justice*: the principle that all persons are of equal moral worth and deserve comparable consideration when moral problems arise. To these some ethicists, such as Sisela Bok, would add the principles of equality and veracity, but I find equality to be a part of the principle of justice, and veracity to be a requirement in acknowledging autonomy (2).

Although not always clearly articulated, these principles infuse modern medical ethics. For example, the first of 10 Principles of Medical Ethics adopted by the American Medical Association in 1957 states that 'The principle objective of the medical profession is to render service to humanity with full respect for the dignity of man. Physicians should merit the confidence of

Growth Hormone Treatment in Down's Syndrome. Edited by S. Castells and K.E. Wisniewski

patients entrusted to their care, rendering to each a full measure of service and devotion' (3). This statement embodies the principles of autonomy and beneficence.

Although it has long been regarded as a distinctive discipline, I adhere to the views of Paul Ramsay that medical ethics is consonant with the ethics of a wider human community, that 'the moral requirements governing the relations of physicians to patients and researcher to subjects are only a special case of the moral requirements governing any relation between man and man' (4). In this chapter I will explore ethical issues that pertain both to physicians and patients and parents and children.

WHY INQUIRE ABOUT THE USE OF HUMAN GROWTH HORMONE IN DOWN'S SYNDROME?

Why are we considering the ethics of using hormone growth therapy in persons with Down's syndrome (DS)? The very act of inviting this analysis suggests a degree of uncertainty, as though one were hoping that an ethical imprimatur might dispel lingering concerns over medical adventurism or provide reassurance about parental motivations.

Why is this topic unsettling? For at least five reasons. The issue challenges us to decide whether a specific feature of DS—short stature—is abnormal (or harms the individual) and, therefore, pursuant to traditional clinical medicine, qualifies for treatment. It demands that we decide whether or not human growth hormone (hGH) is an appropriate therapeutic agent for a person with DS. As with any treatment decision, it requires that a thorough risk–benefit analysis be conducted, and that a decision to treat be preceded by a finding of likely benefit to the individual. It is complicated by the fact that few children with DS have the cognitive power to confront and decide the choice as to whether to undergo treatment with hGH. This necessitates that we consider the motivation of the parent or other third-party adult guardian in reaching a decision on behalf of the child to seek such treatments. Finally, it raises questions about allocation of resources. Is this expensive undertaking worthwhile or should the dollars be spent elsewhere or not at all?

SHORT STATURE, DISEASE AND TREATMENT

It is not possible to draw a line that demarcates health from illness, normal from abnormal. We agree that a woman with breast cancer or a man with congestive heart failure has a serious illness, though on a given day it is quite possible that each feels better than an otherwise healthy person with an acute, transient sinusitis. It is exceedingly difficult to achieve consensus concerning the characterization of other common conditions. Is infertility a disorder that should be treated? What if the person has merely failed to conceive for one year? Is male-pattern baldness a disorder that merits treatment? Should costly medical resources be devoted to reduction mammoplasties or breast augmentations? Such questions are endless.

Is short stature (defined here as more than two standard deviations below the mean for age and sex) a condition that merits treatment? Why or why not? There are many causes of short stature, ranging from primary GH deficiency to growth delay associated with chronic diseases such as cystic fibrosis to constitutional growth delay—a variation in the rate at which a child reaches an ultimately normal adult height. These myriad disorders of growth form a continuum of sorts on which we can plot current thinking about treatment.

When short stature is *severe* (defined here as more than three standard deviations below the mean for age and sex) there seems to be a consensus that treatable causes should be treated, as is evident from the history of the management of girls with Turner's syndrome and children with hypopituitarism. For more than 20 years girls with Turner's syndrome have been treated with oxandrolone to increase final adult height. Recently, the combination of oxandrolone and GH has been shown to yield dramatic (9.8 ± 1.4 cm year) increases in growth velocity in girls between the ages of 10 and 12 (5). A recent report of a multi-center three-year study conducted in Japan of 46 girls with Turner's syndrome has demonstrated a less dramatic, but still impressive, impact of hGH alone. Prior to the development of recombinant DNA technology great efforts were undertaken to harvest GH from cadavers to use in treating children with primary GH deficiency (hypopituitarism). This therapy ended abruptly in 1985 when five children developed the fatal viral disorder Creutzfeld–Jakob disease—a risk that had not been anticipated (6). Fortunately, just a few months later the first genetically engineered hGH product, Proptropin, a drug that today is widely used, received Food and Drug Administration (FDA) approval.

Currently, the major ethical debate in the treatment of short stature is whether otherwise healthy children with constitutional growth delay or familial short stature should be treated with drugs to increase their final adult height. For many years some physicians have been treating boys with constitutional growth delay with low-dose testosterone, and achieving good results as reflected in final adult height (7). But even the more optimistic physicians seem to stop short of advocating the therapy. Why? Because they do not unequivocally define short stature as an abnormality that necessitates treatment. Most require that the patient also demonstrate clear signs of psychosocial damage (low self-esteem, poor socialization skills) or behaviors that indicate such damage is imminent (8). How much do we know about the impact of growth delay and short stature on the child? Relatively little. Although commentators are quick to remind us that there is a height bias in our society (9), there have not been many studies of the impact of unusually short stature on children's sense of self-esteem.

The research literature does suggest that some children with short stature, especially those with certain well-defined causes, suffer significant problems with self-esteem and socialization. One study found that patients with primary hypopituitarism had a 'marked delay in psychological maturation and a lack of aggressive drive'. In contrast, patients with Turner's syndrome were described as normal and persons with achondroplasia were characterized as outgoing, cheerful and optimistic (10). Several studies of children with hypopituitarism have also found a lack of aggression, but did not find the delay in psychological maturation (11). Studies of otherwise normal children with constitutional growth delay have not discerned abnormalities (12). A recent review concluded that there is 'no conclusive evidence that short stature *per se* is associated with psychopathology' (13).

Few studies have addressed the relationship between height and academic performance. One study of 20 children with constitutional growth delay could find no differences across 15 measures when compared to an appropriate control group (14). There is no conclusive evidence that short stature itself is associated with poor school performance. Delayed onset of puberty in boys, a feature of constitutional growth delay, has been consistently associated with lack of independence and low self-esteem. There is some evidence that persons with primary hypopituitarism have life-long problems with psychosocial adjustment. It does not appear that in general children with constitutional growth delay or genetic short stature have these problems or benefit from growth enhancement therapy, although it is certainly true that some otherwise normal short children do suffer from poor self-esteem and would benefit from GH therapy. There is to my knowledge no study that examines the self esteem or socialization skills of children with DS in relation to height or growth rate.

SHORT STATURE, DOWN'S SYNDROME AND TREATMENT

The fundamental clinical problem in DS, the feature that most sharply differentiates the affected individual from normal, is cognitive deficit. Neither stature nor growth rate is emphasized as a significant clinical problem in the leading medical texts that discuss DS. Indeed, the message is relatively reassuring. For example, one authority writes only that 'Growth is relatively slow, and during the first eight years, secondary centers of ossification are often late in development. However, during later childhood the osseous maturation is more normal, and final height is usually attained around 15 years of age' (15). As recently as 1988, an entire volume devoted to the problems faced by young persons with DS did not discuss psychosocial problems caused by growth delay or short stature, nor did it mention therapeutic interventions to increase final height (16).

How different from normal are the linear growth patterns and ultimate stature of persons with DS? According to one recent study, there is about a

'20% reduction in growth rate in infancy (3–36 months) in both sexes, only a 5% reduction in growth rate of girls between 3 and 10 years and about a 10% reduction in boys between 3 and 12 years of age. A more marked reduction in growth rates occurs in adolescent years, i.e. a 27% decrease in girls age 10 to 17 years and a 50% reduction in boys age 12 to 17 years indicating a decreased adolescent growth spurt in Down's syndrome' (17). This is significant, but not severe, and probably occupies a middle ground among disorders associated with impaired growth. At least some DS researchers agree. Reviewing growth data on 382 individuals with DS an Italian group concluded, 'These data seem to demonstrate that growth impairment in DS is not severe, because most of the mean values are at or above 2 SD [standard deviations]' (18). A new book devoted to the medicalcare of persons with DS does consider growth enhancement therapy and concludes that 'it cannot be emphasized too strongly that further controlled and pilot studies need to be carried out before treatment with this potent hormone is administered to Down's syndrome children. It is unknown whether any improvement in eventual height or mental function would occur from early long term administration of growth hormone. Also unknown is the possibility of long-term, adverse effects of administration of growth hormone to Down's syndrome children' (19).

What evidence is there that short stature in DS compromises the individual's ability to seek to maximize his autonomy? I know of no published studies that address this question. Furthermore, it is unlikely that they will be undertaken on behalf of the hundreds of thousands of children with constitutional growth delay, let alone the thousands with DS. The cost of a double-blind long-term, longitudinal study to measure the impact of normalized growth on a variety of psychosocial parameters in children with DS would be great and the return uncertain.

APPROACHING THE QUESTION

How then can we decide whether use of hGH in DS is appropriate? Of course, this requires that we consider the possible benefits and the possible risks, but first we must agree as to the treatment goal. I assume it is to enhance linear growth rate and increase final adult height. There are, of course, other possible goals. There is evidence that early use of hGH increases cranial size in persons with DS—a provocative finding as it may suggest to the lay person the possibility of neurological benefit. But it is only a hint. There is as yet no basis to suppose that therapy with hGH directly affects neurological development. Indeed, it is unlikely that hGH could cross the blood–brain barrier. It is also theoretically possible that long-term hGH therapy could enhance secondary sexual development and, perhaps, improve fertility, arguably laudable goals for patients with DS. GH therapy may reduce the tendency to obesity which negatively affects how some normal persons perceive those with DS. This too is probably a rational goal. It may also help parents to believe that they are doing everything possible for their child. Nevertheless, my focus is on the use of hGH to enhance the adolescent growth spurt and increase final height.

There may be sufficient evidence from the evaluation of children with severe short stature who are not cognitively impaired, such as girls with Turner's syndrome, to infer that the self-esteem of children with DS would be positively influenced by growth enhancement therapy. Unfortunately, none of those studies involved children with significant cognitive deficit, so the inference is less strong than it might be. Nevertheless, assuming that enhanced linear growth rates and final height are reasonable goals, we must decide whether the risks are not so burdensome as to outweigh the benefits.

RISKS

How much clinical experience have we with the provision of hGH to persons with short stature? What risks have been identified? We have more than a decade of clinical experience with the provision of hGH and seven years with its recombinant form, most often in children with primary GH deficiency. The risk of serious side effects appears to be small. Indeed, compared to many drugs it is quite tolerable. Many children with primary hGH deficiency who are treated with hGH develop antibodies but they are (with rare exceptions) not neutralizing and do not impair the growth response. The apparent safety of the drug does not permit us to assume with confidence that the effects of the drug on persons with DS will be congruent. Particularly troubling is the observation that several cases of leukemia have developed in persons treated with hGH for primary GH deficiency (20). Since a number of studies suggest that persons with DS face a 1% lifetime risk of leukemia, the concern that hGH may trigger the disorder cannot be dismissed. Unfortunately, it is simply not feasible to design a realistic method to assess this risk in patients with DS. A very large study lasting decades would be required to determine if the provision of hGH to children with DS would triple the risk of leukemia.

Clinical experience with the use of hGH in DS is growing, but is still quite limited. In a recent study 13 children with DS who were treated with hGH for one year experienced a large increase in linear growth rate (from 5.4 cm per year to 12.2 cm per year) and a significant increase in head circumference. No unusual or disturbing side effects were noted (21). Of course such a brief study permits no conclusions as to the long-term safety of the drug in this population. Even though that hGH appears to be efficacious in enhancing growth in DS there may be special problems with the drug in such patients. Does hGH therapy mask signs of autoimmune disease in treated patients? Could GH therapy that resulted in increased linear growth nevertheless leave parents and/or the child frustrated because the end-point was still far short of expectations? Could dashed expectations harm such parent–child relationships? Does the cost of hGH therapy, which is about $18 000 to $20 000 a year (according to a Genentech representative) and not routinely reimbursed by insurers except when the patient has a primary GH deficiency or Turner's syndrome, harm the family's ability to purchase other important services?

BENEFITS

Assuming that long-term administration of hGH to patients with DS is relatively safe and effective, we face the challenge of weighing its benefits. Let us assume that routine provision of modest doses of hGH could bring most children with DS to a final adult height of one to two standard deviations below the mean—well within the bounds of normal height. Assume also that it brings the head circumference closer to the normal range. Finally, let us also assume that it causes a striking reduction in the tendency to be overweight. That is, the treated adolescent and young adult would have a body build and head size much closer to normal than persons with DS who did not receive hGH. Does this end-point constitute a benefit or set of benefits that justifies the cost and inconvenience (thrice-weekly injections) of the medication? Do the potential gains in self-esteem and social acceptance that could flow from treatment outweigh the apparently small but potential physical risks, the inconveniences, and the high cost of therapy? If it were determined that a rise in self-esteem did not flow from growth enhancement therapy, would it nevertheless constitute a worthwhile intervention?

In seeking the answer to this question, we must weigh intangibles. It is probably true that persons with cognitive deficits who also differ significantly in physical features from normals are less accepted in society than persons who are burdened with comparable cognitive deficits, but have no obvious physical differences. At the risk of sounding cynical, I point out that Christopher Burke, the young man with DS who stars in *Life Goes On* (ABC TV), is burdened with relatively mild features of the syndrome. Being more mildly affected than most persons with DS he helps make the syndrome more acceptable. He also raises our expectations about persons with DS—a situation which can cause both positive and negative outcomes.

SELF-ESTEEM

As Maslow pointed out, 'All people in our society (with a few pathological exceptions) have a need or desire for a stable, firmly based, usually high evaluation of themselves, for self-respect, or self-esteem, and for the esteem of others...satisfaction of the self-esteem need leads to feelings of self-confidence, worth, strength, capability, and adequacy, of being useful and necessary in the world. But thwarting of these needs produces feelings of inferiority, of weakness, and of helplessness' (22).

There have not been many empirical studies of self-esteem or how to build self-esteem in people with substantial cognitive impairments. Much of the research has studied the impact on such persons of being placed in public school settings. It is naive to think that the task of assessing and building self-esteem is monolithic among persons with mental retardation. It is, I think, on clinical grounds impossible to know if efforts to build self-esteem among one set of individuals whose cognitive limitation stems from a specific cause will have a comparable impact upon another group whose impairment has a different cause. As for the impact of growth enhancement on self-esteem, I know of no studies comparing two groups of retarded persons who differed in final height, but were matched for other important variables.

Frameworks for enhancing self-esteem in children with DS have been constructed. They encourage independence, autonomy and self-motivation. They urge parents to help children experience success. They provide for challenges that are real and goals that are realistic. They promote assertive behavior and coping. They offer strategies to enhance body image and personal attractiveness (23). They make sense, but they have not been empirically tested.

I think it fair to assert that persons with DS who are not severely retarded might benefit from hGH therapy through increasing the likelihood of social acceptance and enhancing self-esteem. There is, unfortunately, no easy way to quantify the benefit. Probably, it would range from substantial in some children to no gain at all in others, and even to negative impacts on children who respond poorly to the therapy, fear the injections, and lose other opportunities because of the high cost of hGH. I conclude that the possibility of benefit is real, but that the outcome cannot be known prospectively in any child.

In many respects the issues that arise concerning the use of hGH in DS are not unlike those involving normal children with constitutional growth delay. For more than a decade physicians treating children with short stature have felt uncertain about their work. What is true about treating children with constitutional growth delay with testosterone is true for children with DS with hGH today. We need a series of long-term randomized clinical trials to investigate how alternative treatment regimens, including a placebo, affect final height and improve psychological well-being, and whether they have delayed side effects. Unfortunately, such trials are not likely to be funded.

DECIDING THE QUESTION

If we decide that the administration of hGH has a substantial likelihood of increasing growth velocity and final height, improving body image and enhancing self-esteem, which may in turn enlarge the social opportunities available to persons with DS, and that the risks and inconveniences associated with the therapy are such that reasonable persons might choose it, we must still consider the issues of determining *how* it should be decided whether a *particular* child with DS should undergo treatment.

Before addressing this issue, I shall make two observations. Few (if any) children with DS are capable of weighing the risks and benefits of hGH therapy and reaching a reasoned decision. Many parents of children with DS who have witnessed first hand the pain of social ostracism—the burden upon their child (and themselves) of being different—will be favorably disposed to consider a therapy that tends to reshape their child along parameters that more closely approximate society's sense of what is normal. As a parent myself, it is easy for me to comprehend their motivation. However, I cannot shake the lingering uneasiness that this therapy is a response to an illness in society. If an individual's height had no discernible impact on how society responded to him, then hGH for persons with DS would not generate much interest. Although it is not a reason to decline growth enhancement therapy, I fear that the treatment is a capitulation to a widespread bias in favour of height and reinforces a societal attitude that should be instead challenged. I do not mean to imply that either the child or the parent should therefore be disqualified from being actively involved in a consideration of whether to undergo hGH therapy. I do suggest that the parents should carefully question their motivations.

The child with DS will not be the party who raises the issue of treatment with hGH. Although it is possible that a physician could initiate discussion of this matter, in most cases it will be the parents who raise it. The principle of autonomy requires that the child be as fully involved as possible in the decision as to whether or not to undergo this therapy. I can imagine situations where the child's views should be controlling. For example, consider a 12-year-old severely retarded girl with DS who has for years evidenced extreme discomfort with injections. Subjecting that child, who has completed much of her growth, to thrice weekly injections might produce much anxiety and family turmoil and relatively little growth. I can also imagine situations where the child's views should not be controlling. Consider a nine-year-old, mildly to moderately retarded boy who adores the Boston Celtics. If asked by his father if he would like to take medicine to become taller, it is likely that he would instantly agree.

In most states, pursuant to common law or statutes, there are well-established procedures for determining whether a person who is a minor or an adult who lacks the capacity to make informed decisions, should or should not undergo invasive medical procedures. Parents or physicians who are uncomfortable with the administration of hGH to children with DS could, I suppose, petition a probate court to hear this matter. The judge would appoint a guardian *ad litem* to represent the interests of the child, and hear testimony about the risks and benefits of the drug, and about the child. The judge would decide whether he or she needed to retain jurisdiction in this matter. If that decision was affirmative, the judge would ultimately determine whether it was in the best interests of the child to undergo this therapy. This is the procedure followed, for example, when parents seek to have a retarded teenage daughter sterilized.

In 1979, deciding a case involving the right of parents to commit minors to a state institution for the mentally ill, US Supreme Court Chief Justice Burger wrote: 'Our jurisprudence historically has reflected western civilization concepts of the family as a unit with broad parental authority over minor children. The law's concept of the family rests on a presumption that parents possess what a child lacks in maturity, experience, and capacity for judgement required for making life's difficult decisions' (24). There are of course some situations—requests for sterilization, termination of life-support systems, the administration of psychotropic drugs and a few others—that have been determined by the courts to be beyond the scope of parental decision, but these are the exceptions that reinforce an otherwise broad authority.

Is the decision to request hGH therapy on behalf of the child with DS within the boundaries of parental autonomy or beyond them? As the risks known to be associated with hGH appear to be tolerable and the possibility of benefits to the child with DS from undergoing treatment appear to be real, albeit difficult to quantify, I think that this decision ought to be left to the parents, the child and a knowledgeable physician. This is not a matter that requires judicial intervention.

SCARCE RESOURCES AND REIMBURSEMENT

Having determined that a parent may (not ought) request and consent to the use of hGH as a growth enhancement therapy (and assuming that from a medical perspective there are no competing therapies that are as efficacious while being lower in risk and less expensive), it is still necessary to consider the resource problem.

Human growth hormone for DS is a costly intervention that is not recognized by third-party payers as a reimbursable expense. If and until that changes, for most families the cost factor will influence the decision to seek treatment. Although I would aggressively counsel parents not to initiate hGH therapy for a child if one consequence would be depriving the child or other family members of more important services or life experiences, my reflections on autonomy lead me to conclude that this is a matter for the parents to decide.

Perhaps the most challenging issue for me is whether or not society should agree that hGH is an efficacious agent in DS and pay for it. Ethical analysis is conducted in a world where resources are limited. In an era when containment of health care costs is a central political question, this is especially difficult. Currently, in the USA the only approved use for hGH therapy is in children with documented hGH deficiency. DS children do not fall into this group. Reflecting on the process by which the FDA decides upon an individual use for a drug, I conclude that there is still too little information about this use to stimulate the FDA to approve its use. Turning to the issue of coverage, whether by private insurers, health maintenance organizations (HMOs) or government programs, I also conclude that there is not yet a sufficient body of evidence of the positive impact of hGH in DS children to compel payment. Hopefully, such evidence will emerge in the near future.

This may seem harsh. It is not so intended. There are many possibly valuable medical interventions that insurance plans will not cover (e.g. liver transplants, bone marrow transplants for metastatic breast cancer). That list is likely to grow rather than shrink. Ethical analysis may ignore economics, but it then runs the risk of reaching conclusions about what ought to be done that have little likelihood of being implemented.

CONCLUSION

I conclude that the principle of beneficence supports the decision of parents and physicians to consider the administration of hGH to some children with DS. The principle of autonomy requires that so far as possible the views of the child be factored into the decision-making process. Sustained objections to the injections, for example, might constitute grounds for halting therapy. The principles of autonomy and non-maleficeance may require that a physician refuse to comply with a parental request for therapy should he or she decide that it is not in the best interests of the child. The principle of non-maleficeance may require that an FDA official seek more data before deciding whether or not to approve the use of hGH in DS. The principle of justice, which seeks equitable distribution of resources, may demand that hGH therapy for children with DS not be reimbursed absent appropriate studies that demonstrate positive therapeutic effects upon self-esteem and socialization.

REFERENCES

1. Wertz DC, Fletcher JC, Mulvihill JJ. Medical geneticists confront ethical dilemmas: cross-cultural comparisons among 18 nations. *Am J Hum Genet* 1990; **46:** 1200–1213.
2. Bok S. The tools of bioethics. In *Ethics in Medicine*, edited by Reiser SJ, Dyke AT, Curran WJ, MIT Press, Cambridge, MA, 1977; pp. 138–140.
3. American Medical Association. Principles of medical ethics. In *Ethics in Medicine*, edited by Reiser SJ, Dyke AJ, Curran WJ, MIT Press, Cambridge, MA, 1977; pp. 38–39.
4. Ramsay P. *The Patient as a Person*, Yale University Press, New Haven, 1970; p. 1.
5. Blizzard RM, Hindmarsh PC, Stanhope R. Oxandrolone therapy: 25 years experience. *Growth* 1991; **7:** 1–6.

6. Brown P, Gajdusek DC, Gibbs CJ, Asher DM. Potential epidemic of Creutzfeld–Jakob disease from human growth hormone therapy. *N Engl J Med* 1985; **313:** 728–731.
7. Kaplan JG, Moshang T, Bernstein R, Parks JS, Bongiovanni AM. Constitutional delay of growth and development: effects of treatment with androgens. *J Pediatr* 1973; **82:** 38–44.
8. Richman RA, Kirsch LR. Testosterone treatment in adolescent boys with constitutional delay in growth and development. *N Engl J Med* 1988; **319:** 1563–1567.
9. Murray T. The growing danger from gene spliced hormones. *Discover* February 1987; 88–92.
10. Drash PW, Greenberg NG, Money J. Intelligence and personality in four syndromes of dwarfism. In *Human Growth: Body Composition, Cell Growth, Energy, and Intelligence*, Lea & Febiger, Philadelphia, 1968; pp. 568–581.
11. Drotar D, Owens R, Gotthold J. Personality adjustment of children and adolescents with hypopituitarism. *Child Psychiatry Hum Dev* 1980; **11:** 59–66.
12. Gordon M, Crouthamel C. Post EM *et al.* Psychosocial aspects of constitutional short stature: social competence, behaviour problems, self-esteem, and family functioning. *J Pediatr* 1982; **101:** 477–480.
13. Lee PDK, Rosenfeld RG. Psychosocial correlates of short stature and delayed puberty. *Pediatr Clin North Am* 1987; **34:** 851–863.
14. Gordon M, Post EM, Crouthamel C *et al.* Do children with constitutional delay really have more learning problems? *J Learn Dis* 1984; **17:** 291–293.
15. Jones KL. *Smith's Recognizable Patterns of Human Malformation*, 4th edn, Saunders, Philadelphia, 1988; p. 11.
16. Pueschel SM *The Young Person with Down's Syndrome*, Brookes, Baltimore, 1988.
17. Cronk C, Crocker AC, Pueschel SM *et al.* Growth charts for children with Down's syndrome: 1 month to 18 years of age. *Pediatrics* 1988; **81:** 102–110.
18. Piro E, Pennino C, Commarata M *et al.* Growth charts of Down's syndrome in Sicily: evaluation of 382 children 0–14 years of age. *Am J Med Genet* (Suppl.) 1990; **7:** 66–70.
19. McCoy EE. Endocrine function in Down's syndrome. In *Down's Syndrome: Advances in Medical Care*, edited by Lott IT, McCoy EE, Wiley–Liss, New York, 1991; p. 81.
20. Fisher DA, Job J-C, Preece M, Underwood LE. Leukaemia in patients treated with growth hormone. *Lancet* 1988; **i:** 1159–1160.
21. Torrado C, Bastian W, Wisniewski KE, Castells S. Treatment of children with Down's syndrome and growth retardation with recombinant human growth hormone. *J Pediatr* 1991; **119:** 478–483.
22. Maslow AH. *Motivation and Personality*, Harper & Row, New York. 1954.
23. Frank RA. Building self-esteem in persons with Down's syndrome. In *The Young Person with Downs' Syndrome*, edited by Peuschel SM, Brookes, Baltimore, 1988; pp. 205–213.
24. *Parham* v. *J.R.* 442 US 584 (1979).

Discussion of Presentation

PHILIP REILLY
(Edited by D.A. Snider and Salvador Castells)

D. Lawrence: My oldest daughter was born with Turner's syndrome. My second daughter was born with Down's syndrome. The doctor was very quick to help my child with Turner's and nothing to help my child with Down's. A physician in California put her on a metabolic treatment, which seems to have brought about the same results as growth hormone treatment. I wonder about the medical ethics of a doctor being so eager to help a child with Turner's and devoid of information about Down's. I should point out that I wasn't disappointed in my children.

P. Reilly: I agree with you that children who have Down's syndrome should have the same claim to therapy as children who are not cognitively impaired. They have the same moral claim. The hard question that one confronts is, how can we measure self-esteem? Will such treatment improve self-esteem in the more severely impaired? You, as a parent, have every right to request and consent to this therapy on the behalf of your child. If a physician does not wish to proceed with this therapy, he has the duty to refer you to someone who might be able to help you.

20 Family Nursing Intervention for the Down's Syndrome Child in Human Recombinant Growth Hormone Treatment

ARLENE HURWITZ
City University of New York, City College, School of Nursing, New York, USA

INTRODUCTION

Treatment with recombinant human growth hormone (rhGH) can correct growth retardation and other problems associated with Down's syndrome (DS). Becoming familiar with diagnostic methods and treatment with synthetic GH is a prerequisite for nurses caring for families of DS children. Family intervention involves assessment and education of the family members, including injection technique, and ongoing support of the family unit.

This chapter reviews the current research on human recombinant treatment for children with DS and the implications for family nursing intervention. The nursing process is used to systematically outline the steps in assisting DS families with children in experimental treatment.

The results of recent studies focusing on DS children, show that treatment with rhGH results in head growth, greater muscle strength and increased height. These research findings may translate into new nursing strategies for managing the symptoms afflicting children with DS. Increasingly, nurses will participate in diagnostic assessment, therapeutic management and the provision of long-term family support. It appears likely that nurses in the near future, working in conjunction with physicians, will assume the role of case manager for the family of a DS child in experimental treatment.

HEALTH PROBLEMS OF CHILDREN WITH DOWN'S SYNDROME

DS or trisomy 21 is a genetic problem which results in mental retardation, microcephaly, muscle weakness and short stature (1–3). Children born with DS have 47 chromosomes instead of the normal 46. Infants may experience

Growth Hormone Treatment in Down's Syndrome. Edited by S. Castells and K.E. Wisniewski

dry skin, frequent upper respiratory infections, and feeding/eating and perceptual difficulties (4).

Recent scientific inquiry of the treatment of children with DS has brought about the experimental use of synthetic GH (5). Eighteen families with short-stature DS children have elected to have their children treated according to the experimental protocol. To date, the results of the treatment are encouraging and are significant to pediatric nurses working with DS families. The need for nursing intervention with families of children with chronic illness has long been cited in the nursing literature (6, 7). In the case of families with a DS child, nurses play a pivotal role in the coordination of services to the child living at home (8, 9). A review of the literature indicates that immediately following the birth of the child, families of DS children are presented with a complex set of challenges. Initially, they must form a realistic and positive identification with the infant and begin to mourn the infant 'lost' (10). It is not uncommon for family members to be fraught with sadness while dreading the profoundly resultant impact that rearing a DS child will have upon their once familiar domestic realm (11, 12). Fathers have been found to withdraw from caregiving and tend to retreat from their friends. Mothers have reported feelings of sadness that vacillate over time from mild to severe (11). The intervention of nurses has helped both to neutralize many of these feelings and to create a more productive family support system.

NURSING PROCESS: ASSESSMENT

Assessing the DS family with a child in or considering experimental treatment is similar to assessing DS families in general, inasmuch as the structure, development and function of the family are appraised (13). On the other hand, it is important to know the type and source of information the family already has about the experimental treatment to address any of their questions and to dispel any misinformation. In addition, consideration should be given to the family's expectations and to their monetary resources and community networks.

In conjunction with the overall family assessment, the DS child's individual assessment takes place during two different clinic visits. The first of these visits entails completing a family history, genetic profile, 24 h diet recall, a urine analysis and a comprehensive physical examination. The child is weighed and measured and the cranial circumference is obtained. An 8 h fasting venous blood sample is drawn for SMA-12 and 6, Eastman Kodak 700 analyzer (Eastman Kodak Co., Rochester, NY) to perform complete blood count, blood lead level, thyroxine somatomedin-C concentrations, and other tests.

The second clinic visit is comprised in part of reviewing of the laboratory results and gathering additional data on GH concentrations. Because DS children have been reported to have an increased incidence of hypothyroidism (3, 14), special attention is paid to the thyroid function test. The procedure for gathering information in serum GH concentrations consists of a series of blood samples taken from a venous catheter during the administration of L-dopa and clonidine—two drugs that stimulate the hypothalamus to produce GH-releasing hormone, which in turn stimulates the pituitary to produce GH (15–17). Serum concentrations of GH are measured in plasma to determine if it is a deficiency. Since serum GH concentrations can fluctuate, a single plasma sample concentration of GH is not diagnostic of GH deficiency; therefore a series of blood samples after a stipulatory test is necessary. L-dopa is administered orally according to body weight: if less than 15 kg, 125 mg is given, and between 15 and 30 kg, 250 mg. Blood samples are drawn at 20, 40, 60, 80, 90 and 120 min after L-dopa administration. Clonidine is administered orally as a single dose of 0.15 mg/m^2 body surface area. Blood samples for determination of serum GH concentrations are drawn at 0, 30, 60, 90 and 120 min. The child is considered GH deficient if serum concentrations of GH do not increase over 10 ng/ml after the administered tests. DS children seem to have a normal pituitary function as production of GH is concerned (18), but a deficient control by the hypothalamus, indicating a classic GH neurosecretory dysfunction (19). If the DS child is found to be GH deficient, methionyl rhGH, 0.1 mg/kg, is administered three times a week subcutaneously. The drug is administered by one parent at home after one to two weeks of outpatient instructions and training by the nurse.

NURSING DIAGNOSIS, PLANNING, IMPLEMENTATION AND EDUCATION

The nursing diagnosis relates to the family's knowledge of the GH deficit, and it is specially related to the experimental treatment protocol and the anxiety about the unknown outcome of the experimental treatment (20).

Planning for both the child and family involves consideration for the child's emotional and physical safety, the parent's or caregiver's participation in assisting the child during the diagnostic testing, and the side effects of the treatment's medications. Planning for follow-up evaluations of the continuous home treatment is also essential in monitoring its outcome.

Diversion and security for the child can be achieved if parents are advised, prior to testing, to bring one of the child's favorite toys or books. The family, including the DS child, should be given appropriate explanations pertaining to the procedures of the diagnostic protocol. In the course of the first clinic visit the nurse has an opportunity to provide reassurance to the child by listening to the child's concerns and by assisting the parents in supporting the child. Documented side effects of the administration of L-dopa include dizziness and sleepiness. The nurse and family, in anticipation of these possible side effects, can provide assistance such as periods of rest for the child or reading aloud to the child. Additionally, the child's blood pressure should be monitored closely because of the hypotensive effects of clonidine. During the second clinic visit resources or referrals that might be appropriate such as information about pica are discussed with the parents.

Once given the laboratory results, the parents are faced with the decision of whether or not to enter their child into treatment. During this time of decision, the nurse acts as a neutral support assisting the family in making an informed choice.

In the instance that the parents opt to accept treatment for their child, the nurse and family plan continuous weekly treatment. Instructions are given to the family in preparation for their in-home administration of the GH to their child. Subcutaneous injections of synthetic GH are administered three times per week. Information given to parents concerns sterile technique, proper positioning of the child, and selection and rotation of the injection sites. Parents must be supported to work through the apprehensions and guilt associated with giving injections to their child (21).

The nurse can evaluate the progress of the parent/caregiver in the administration of the GH and can determine the parent's ability to give injections that might be difficult to perform on their own children. Ongoing evaluation can be accomplished, on a monthly basis, either by home visits or phone calls.

CONCLUSION

The increase in height and head circumference produced by long-term administration of hGH in DS children underscores the efficacy of this treatment in lessening growth retardation. This alone may have a substantial impact on the self-image of DS children. Administration of GH therapy in conjunction with other innovative treatments, such as reconstructive surgery and early cognitive therapy, may substantially reduce the social impact of this chromosomal abnormality on these children and their families. It is hoped that DS children will achieve better social integration if their height and appearance are closer to those of non-affected children. Further, improved coordination and physical abilities may contribute to enhanced performances in all activities, further building hope and confidence. As techniques improve, DS children may well become capable of integration into mainstream schools and society at large and go on to lead fulfilling and productive lives. Pediatric nurse practitioners will be faced with new challenges as they assume the role of case managers for DS families and take greater responsibility for provision of new treatments such as hGH therapy.

REFERENCES

1. Wisniewski KE. Down's syndrome children often have brain growth with maturation delay, retardation of growth and cortical dysgenesis. *Am J Med Genet* 1990; **7:** 274–281.
2. Cronk C, Crocker AC, Peuschel SM *et al.* Growth charts of children with Down's syndrome: 1 month to 18 years of age. *Pediatrics* 1988; **81:** 102–110.
3. Peuschel SM. Growth, thyroid function, and sexual maturation in Down's syndrome. *Growth Genet Horm* 1990; **6:** 1–5.
4. Pueschel S, Rynders JE. *Advances in Biomedicine and the Behavioral Sciences*, Ware Press, Cambridge, MA, 1982.
5. Castells S, Torrado C, Bastian W, Wisniewski KE. Growth hormone deficiency in Down's syndrome children. *J Intell Disab Res* 1992; **36:** 29–43.
6. Hymovich DP, Chamberlin RW. *Child and Family Development Implications for Primary Health Care*, McGraw-Hill, New York, 1980.
7. Friedman MM. *Family Nursing: Theory and Assessment*, Appleton-Century Crofts, E. Norwalk, C, 1981.
8. Russell FF. Intervening with families of infants with Down's syndrome. In *Families and Chronic Illness*, edited by Wright L, Leahey M, Springhouse Publishing, Springhouse, PA, 1987; pp. 257–274.
9. Miola ES. Down's syndrome: update for practitioners. *Pediatr Nurs* 1987; **13:** 233–237.
10. Rommey M. Congenital defects: Implications for family development and parenting. *Issues Compr Pediatr Nurs* 1984; **7:** 15.
11. Damrosh SP, Perry LA. Self-reported adjustment, chronic sorrow, and coping of parents of children with Down's syndrome. *Nurs Res* 1989; **38:** 25–30.
12. Fewell R. A handicapped child in the family. In *Families of Handicapped Children*, edited by Fewell R, Vadasy P, Pro-Ed, Austin, 1986; 3–34.
13. Wright L, Leahy M. *Families and Chronic Illness*, Springhouse Publishers, Springhouse, PA, 1987.
14. Torrado C, Bastian W, Wisniewski KE, Castells S. Treatment of children with Down's syndrome and growth retardation with recombinant human growth hormone. *J Pediatr* 1991; **119:** 478–483.
15. Tapanainen P, Knip M, Lautala P, Leppäluoto J. Variable plasma growth hormone (GH)-releasing hormone and GH responses to clonidine, L-dopa, and insulin in normal men. *J Clin Endocrinol Metab* 1988; **67:** 845–849.
16. Gil-Ad I, Topper E, Laron Z. Oral clonidine as a growth hormone stimulation test. *Lancet* 1979; **ii:** 278–279.
17. Lanes R, Hurtado E. Oral clonidine: an effective growth hormone-releasing agent in prepubertal subjects. *J Pediatr* 1982; **100:** 710–714.
18. Castells S, Beaulieu I, Torrado C, Wisniewski KE, Zarny S, Gelato MC. Hypothalamic versus pituitary dysfunction in Down's syndrome (DS) as cause for growth retardation. International Conference on *Growth Hormone Treatment in Down's Syndrome*, New York, 10–11 October 1992 (abstract).
19. Spilliotis BE, August GP, Hung W, Sonis W, Mendelson W, Bercu BB. Growth hormone neurosecretory dysfunction, a treatable cause of short stature. *JAMA* 1984; **251:** 2223–2230.
20. Hurwitz A. Helping Down's syndrome children to reach new heights: the role of the pediatric nurse. *Pediatr Nurs* 1992; **18:** 485–489.
21. Coody D, Stevens B. Nursing strategies: altered endocrine function. In *Family Centered Nursing Care of Children*, edited by Foster RLR, Hunsberger MM, Tackett Anderson JJ, Saunders, Philadelphia, 1989; pp. 1898–1956

Discussion of Presentation

ARLENE HURWITZ
(Edited by D.A. Snider and Salvador Castells)

Z. Laron: Who performs these tasks in the United States and who provides approval when a private physician prescribed such therapy?

A. Hurwitz: In this country, the nurse acts as case manager for children with chronic illnesses and families that require long-term support. Giving injections is in the realm of a nurse, not a social worker.

S. Pueschel: Regarding Dr Laron's second question, a private physician does not need approval to provide such therapy, if the drug is approved by the Food and Drug Administration.

S. Castells: We work in two different health science centers. One is a pediatric service—the SUNY Health Science Center at Brooklyn—and the other is a neuroendocrine clinic at the New York State Institute for Basic Research in Developmental Disabilities on Staten Island. We have human experimentation committees in each institution, which have approved our protocol. However, it is my understanding that physicians can decide treatment on their own since rhGH has been approved by the Food and Drug Administration for use in GH deficient children.

P. Reilly: Dr Pueschel and Dr Castells have interpreted correctly the use of rhGH by practitioners. Actually the laws in the USA have little to say on this question. There is one problem that physicians can have in prescribing essentially experimental therapeutic agents: should there be an untoward event that becomes a factor in the malpractice case that follows.

G. Annerén: I think there is a big ethical problem in GH treatment in Down's syndrome, and it can't be compared with the ethical problem in Turner's syndrome where the effects are known and there have been extensive trials. We don't know anything yet about the final height in Down's syndrome. We don't know anything yet about brain growth and about intellectual capacity. We know they grow during the treatment, but I don't think GH treatment should be done except within controlled studies.

21 Family Perspectives on Growth Hormone Treatment in Down's Syndrome

DENISE L. DUFFEY
Central Florida Down Syndrome Association, Inc., Orlando, Florida, USA

While we were making the film for this conference, I asked the producers to put more family scenes in, because I noticed that when a family was shown I saw, not a child with Down's syndrome, but a family. And that is exactly what we see when we look at our children with Down's syndrome, a member of our family. If you found yourself doing that, too, then you begin to understand the real family perspective on any medical treatment for our children. Because of the extra efforts required to keep them healthy and learning, these children are usually the most appreciated and cherished members of our families, really the 'love of our life'.

Because we love our children so much we truly appreciate the important research that the presenters at this conference have done in an effort to help our children and would like to thank them all for their hard work.

By the time our children are a year old, we have been through so much with them medically. We learn quickly what a difference good medical care makes in their development. Early detection and aggressive treatment of the more frequently occurring conditions such as hearing impairment, vision impairment and thyroid dysfunction have kept many of our children from being impaired by a secondary disability.

Comprehensive heath care, proper treatment of infections and congenital heart disease, as well as the impact of early intervention and inclusion in the educational system, all have improved the quality of life of children with Down's syndrome today, allowing them to live longer and actively participate in our families, our schools and our communities (1).

Most parents feel that good, basic medical care is the most critical factor in their child's development in the first years. Yet most parents find themselves at odds with the medical establishment from the time their child is born.

Few of us are happy with the way our doctors, nurses or our social workers handled the birth of our child. Most of us did not receive any printed or current information about Down's syndrome. Our doctors come in to point out the physical defects that indicate our new babies have Down's

Growth Hormone Treatment in Down's Syndrome. Edited by S. Castells and K.E. Wisniewski

syndrome, but offer little information about early intervention or health care for them.

We go to our pediatricians hoping to have our child treated like they have a real medical condition, but usually are told there is nothing special that needs to be done. We end up switching pediatricians several times in the hope of finding someone whom we could feel reassured would be alert to health problems associated with Down's syndrome and help us avoid the many conditions that could impede their development.

We begin to 'network' with other parents, trying to get information about anything they are doing that our pediatricians may not know about. When my child was born, many of the parents in Orlando were using illegal 'sheep cell therapy' from a doctor in Germany. Numerous parents swore their children were experiencing immediate and noticeable benefits from this treatment. When we would go to our pediatricians for help in evaluating this type of treatment, they would either say it was ridiculous or 'go ahead and try it; maybe it will do something'. They did not bother to help us get the research data or information we needed to protect ourselves against harmful treatments.

When we decided to form a new parents' association last year, the first two meetings consisted mainly of complaints about doctors and the medical system. We heard complaints about the dentist who told a mother she needn't worry about her son's missing teeth, because he was 'a mongoloid, after all' and his appearance did not matter. Or the little baby who almost went into congestive heart failure at three months, because his doctor did not tell his parents that he had two holes in his heart.

The message we received time and again from other parents was 'you won't get any help from the medical establishment. They don't care about children with Down's syndrome'. It was not just a few isolated instances; it was the norm. We heard the same complaints from parents all over the country.

Just last month we heard a dentist on a medical panel at the National Down's Syndrome Congress Conference in Atlanta tell a mother that he would not normally recommend braces for children with Down's syndrome. He suggested that if you close the gaps in their teeth, they would develop gum disease. The parents in the audience shook their heads and we all wondered if he did not know that our children with Down's syndrome brush their teeth, just like their brothers and sisters do. Where did he get his picture of Down's syndrome?

After the article on growth hormone treatment for children with Down's syndrome was published in the *Journal of Pediatrics* last year (1999), Dr Castells received many letters from doctors saying that it was 'wrong to waste research money on children who are retarded' and that it was 'not important for these children to grow taller and look nicer'.

Both parents and doctors working with children with Down's syndrome wonder where on earth these attitudes keep coming from. Why would any doctor think so little of our children that they would suggest withholding basic medical treatment from them?

But as I have worked the last few months on this speech, I realized something that I had not before. I realized that most parents with young children like mine have almost no concept of what Down's syndrome was like 25 years ago, yet for most in the medical profession this experience is what their concept of Down's syndrome is still based on today.

I kept thinking about the statement that as recently as 1965, 50% of children with Down's syndrome were dead by age five, due mostly to unrepaired congenital heart disease, and tried to picture what life for a child with Down's syndrome and their families must have been like then. Surely nobody was concerned about developing early intervention programs. The idea of mainstreaming was only a dream for most families even 10 years ago, and total inclusion in the regular classroom for all children is an ideal that is only beginning to be considered today.

What a difference basic medical treatment has made in the potential for children with Down's syndrome and how much it must have changed our expectations for them!

My impression is that the average health care professional we are working with today, including those who wrote the letters to Dr Castells, is simply not aware of the changes taking place to help children with Down's syndrome, just as we parents have trouble visualizing what Down's syndrome was like in the past.

I do not believe that the average pediatrician really comprehends the benefits of early intervention programs, or parents would not have so much trouble getting prescriptions for therapies. I do not think they understand the joy and lessons about life that our families experience when they raise a child with special needs and how they are real members of our families, or doctors would be more positive when our children are born. And I do not think they realize, as we do, that aggressive attention to every health issue is critical, making the greatest difference in our children's quality of life and potential, or they would be more aggressive about our children's medical treatment.

The truth is that so many dramatic changes have come about in the last 25 years, 10 years, and even in the last year for children with Down's syndrome that it is very hard for any of us—parents, doctors, or even the national Down's syndrome organizations—to keep up with all of the changes fast enough to get help to our children as quickly as it is needed.

I saw an example of this at the National Down's Syndrome Congress Conference last month in Atlanta. A wonderful speaker, Mr Norman Kunc, who has cerebral palsy and had been in 'special schools' and special education classes throughout his youth, spoke out very strongly against any child being segregated into special education classrooms or schools. He explained to us that we must begin to demand total inclusion in the normal classroom for *all* children with disabilities (2).

He stated that no person, regardless of the extent of his disabilities, should be segregated and have to earn the right to be a member of our society. Like many others today, he feels it is more important for our children to be able to attend their neighborhood school with their brothers and sisters, learning how to live in the real world, than it is to sit in a special education classroom all their lives, only to progress to a sheltered workshop in adulthood.

Many of us with younger children were thrilled to hear this concept being promoted by the Congress. The parents who own the school where our association's meetings are held and whose five-year-old son has Down's syndrome are already implementing a total inclusion program at their school. They have seen all of their school children benefit from this type of program and we have been watching their progress with great excitement.

I noticed, however, that several of the parents of children 10 years old or so who were sitting in front of me during Mr Kunc's speech were not so delighted with this idea. Here they had struggled to find decent 'special schools' for their children, where they could get therapy and learn basic skills every day in an environment where their children would be loved and sheltered. Now they were being told that it is no longer the acceptable thing to do, that such settings are actually harmful to their children's development. What a tremendous change in just the last couple of years!

Indeed the whole concept of mental retardation itself is being challenged, or at least redefined. Dr Krystyna Wisniewski, a neurologist and one of the researchers on this project, has felt for several years that many of our children with Down's syndrome will be only 'learning disabled' or mildly retarded if they receive the benefits of early intervention (3).

Indeed, today our parents are working very hard to diagnose and treat the learning disabilities and behavior problems in their children with Down's syndrome just as they would any other child, with noticeable results. Therapies based on building neurological development have been so successful in our early intervention programs. Yet parents are still being told by the national Down's syndrome organizations that our children will, almost without exception, be retarded, which sets the stage for lowered expectations from everyone.

Another speaker at the Congress Conference challenged the concept of mental retardation even further. According to Dr Anne Donnellan PhD, a professor of psychology and special education with the University at Wisconsin in Madison, in view of recent research, 'we cannot even begin to define metal retardation'. In a stunning address to the plenary session at the National Down's Syndrome Congress in Atlanta last month, Dr Donnellan told us about research being conducted with persons with all types of disabilities, including Down's syndrome, using 'facilitated communication' (4).

In facilitated communication, a 'facilitator' actually holds the arm of a disabled person to enable them to hit the keys they want to hit on a keyboard. The researchers learned that many persons who are autistic or otherwise mentally or physically disabled and had been unable to communicate with us before were actually able to read and write and quite aware of what was happening around them. But they could type on the keyboard only with physical assistance. The researchers were surprised to find that persons with Down's syndrome also needed this type of physical assistance to type.

In her workshop, we saw a film of a 35-year-old man with Down's syndrome, whom they assume had never been in early intervention and whom we probably would not think was very physically impaired, although his speech was very poor. He convinced them to let him try the keyboard and was surprisingly able to type out messages, such as 'Try this with people who look like me'.

I talked with a parent last week who has used this type of facilitated communication with her 10-year-old son. His condition is undiagnosed but he has many of the same symptoms as Down's syndrome. She said his teachers at school, in trying to test his comprehension, would ask him questions such as 'Where is the floor?' They wanted him, of course, to at least point at it, but he did not. Later, sitting at a keyboard with his mother holding his arm, when asked where the floor was, he typed 'Underneath the table'.

At the end of the workshop, a colleague of Dr Donnellan noted that they had not told us even half of what they were learning, and said that if they did we would not believe them. This type of work begins, I hope, to redefine Down's syndrome and other types of disorders as physical and medical conditions, not just as 'mental retardation' or mental deficiency.

In the plenary session, Dr Donnellan told the Conference that many of these disorders were described over 100 years ago as a 'movement disease', or a neurological condition that prevents rapid processing of information between the brain and the body. My husband and I were shocked when she said this, because it described exactly the changes that we saw in our daughter, Chelsea, just six months after she started growth hormone treatment.

My husband, who is an electronics engineer designing computerized traffic control systems, likens the change in Chelsea to the difference between an old computer and the new, faster microprocessors. She just was able to process faster and respond quicker. Everything was working better and faster.

Indeed, we saw many changes in Chelsea within six months after treatment started. In addition to a height increase of four and a half inches the first year, her head, which had really almost stopped growing, grew another inch. The muscles in her face, lips and tongue suddenly came in, giving us the feeling that we were really seeing our child as she was meant to look for the first time in her life.

Typically for a child with Down's syndrome, she wasn't able to crawl until she was 15 months old. Shortly after treatment started, she began not only to walk, but within six months was able to walk across the room on her tiptoes. She seemed much stronger.

After six months of treatment, she could not only pop eight puzzle pieces right into place, but she could also name them. She had only said about 10 unclear words prior to treatment and had not even been considered ready to start speech therapy until three months after treatment started. Eight months later her speech therapist said we didn't have to come back every week, because 'her speech was developing like that of a normal child'. All of the children with Down's syndrome we know who are Chelsea's age or even older are still having great difficulty with speech and pronunciation.

The changes we saw in Chelsea within six months of treatment were dramatic. Today, Chelsea is a nice-looking little four-year-old who speaks clearly and will be accepted into any school we choose for her. She has never been in any intensive stimulation and education programs; we are happy that she is growing and looks so nice and is able to talk and learn along with normal children instead.

Her teacher last year learned as much as she could about Down's syndrome before getting Chelsea in the classroom. Afterward, she told us she did not see what the problem was with Down's syndrome and that they had had some very interesting conversations.

More than anything else, I was so grateful when at two and a half she was finally able to sing songs instead of sign them. It was so wonderful to have a three-year-old child with Down's syndrome who could ask 'Why?' and wonder where the water in the sink goes, ask about dying, and where babies come from. I love to hear her shrieking with laughter at her own silly knock-knock jokes. These are amazing experiences for parents of a young child with Down's syndrome.

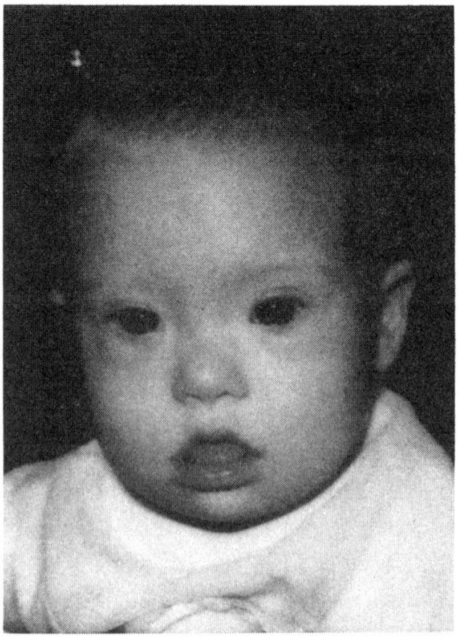

Figure 1. Untreated—12 months

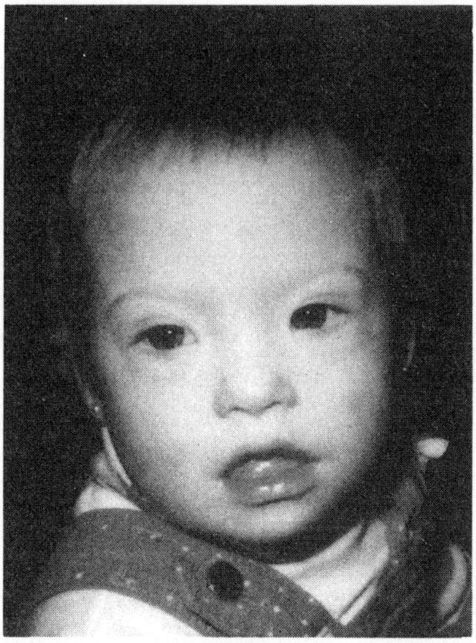

Figure 2. Untreated—18 months

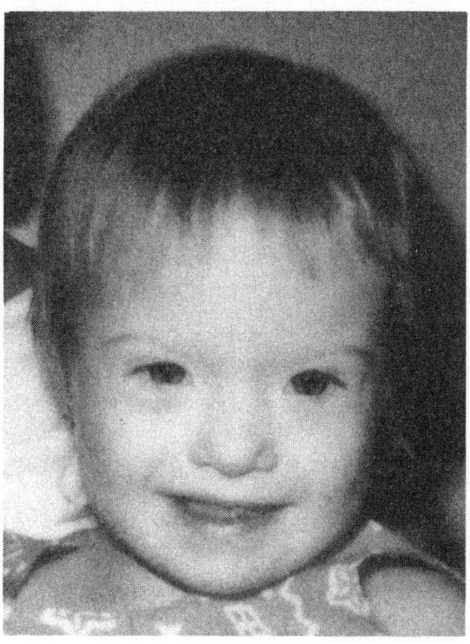

Figure 3. Six months after growth hormone began, to 27 months

Our child was lucky enough to receive treatment for one of the major symptoms of Down's syndrome—improper growth. Although recent testing has shown that her brain has grown well and does not have the structural abnormalities typically found in children with Down's syndrome, we wish that she could have received it much earlier than 20 months of age.

It was reported several years ago that many of the physical features of Down's syndrome, such as small heads and slanted eyes, underdeveloped palates, small ear canals, stubby hands and feet, are directly related to poor growth (4) . Our own personal experience with our daughter and preliminary research findings give us great hope that if these children receive treatment while still infants we will see dramatic improvements not only in their appearance but in their health, speech and neurological development (5). We believe growth hormone treatment will change the face of Down's syndrome forever, literally.

We feel that statements such as the one that 'we should not waste research money on children who are retarded' are being made by doctors why simply do not know how much our children are benefiting from the help they are receiving today. They are unaware that the whole concept of Down's syndrome and even mental retardation itself are changing so dramatically. They are certainly not aware of the promising benefits of brain growth and improved speech that preliminary research with growth hormone treatment is showing.

Indeed, so much money is currently being spent on special education and medical needs for our children that any research money spent to alleviate their condition is certainly money well spent. It is a great emotional and financial stress on the average family to have a child who requires so much therapy, medical care and stimulation. We do not 'just take them home and love them' any more. It is a burden we gladly accept because we know that what we are doing is making a real difference for them, but if any help in alleviating this burden is available, it should not be withheld.

I would like to answer some of the other charges critics of medical treatment for children with Down's syndrome make. As the idea of 'inclusion' in the school system spreads, height and appearance are going to be of even more importance to children with Down's syndrome and their families than ever before, as a story told to me by a therapist who moved to Orlando last year illustrates.

She told me about her lovely 14-year-old sister with Down's syndrome, who went through one of the best early intervention programs in the country. She was mainstreamed into the normal classroom beginning in preschool and did very well. But as she grew older, the teasing and prejudice she experienced as a result of her obvious Down's syndrome appearance and unclear speech caused her such pain she became very depressed and began missing school. She finally refused to go to school altogether, and actually requested that she be allowed to go to a special school.

The problems she experienced in continuing in the normal classroom were not because of any deficiency in her schoolwork, or because of any behavior problems or handicaps. They were only because her appearance and unclear speech marked her as 'one of those retarded kids'.

Another family in Jacksonville, whose 11-year-old son is still only the size of a second grader, worried how he would ever be able to get a job because he is so short. Their pediatric endocrinologist did not want to test him for growth hormone deficiency, saying 'he does not know or care that he is short'. His parents know he does.

One noted doctor worries 'If our kids grow taller and look nicer, won't people expect too much of them?' What if they expect too little of them? We are certainly expecting an awful lot of them, and ourselves, now. Is that not why to a large extent, they are doing so well today, because our expectations for them are so high?

Why then, after all the hours of caring, love and hard work, would we want to handicap them with the appearance of mental retardation, if it could be helped? This doctor is, unfortunately, basing his view on children who have not received the benefits of growth hormone in infancy, when it might also improve their brain growth and speech, according to preliminary research indications.

A friend recently sent me a newspaper article about a physicians' committee and 'watchdog group' who are trying to get the National Institutes of Health to stop all growth hormone testing on any children, because they feel it is cruel to give children three shots a week just to improve their height and appearance. Would these same doctors suggest that it is cruel to give diabetic children insulin injections every day, or to administer thyroid medication to children who need it?

The fact is that growth hormone is not an issue of height and appearance, but one of medical necessity, just as these other treatments are. My husband and I did not seek growth hormone treatment for our daughter because she was short, but because she was not growing normally during the important infant stage. We were especially concerned that her head growth had stopped, because we were told 85% of brain growth occurs in the first year.

Everything in our body needs to grow during that critical time—our brains, our neurological systems, and organs. When we looked at the Physician's Desk Reference and saw all of the systems that were affected by growth hormone deficiency, including metabolism, we concluded that growth hormone treatment was a medical necessity for our child, essential for both her growth and development. We just wanted the same testing and treatment for our child with Down's syndrome that any other child would receive. It seems to us that the effects of growth hormone *deficiency* are much more harmful than any side effects treatment has been shown to have.

We would also remind these critics that the federal Child Abuse Amendment of 1984 provides that *all* infants must under *all* circumstances be provided nutrition, hydration and medication, and *be given any medically indicated treatment* (1). As parents, we believe it is illegal, unethical and immoral to suggest withholding any needed medical treatment from a child based on preconceived, outdated notions of their mental potential.

Because we feel good medical treatment is the most critical need for young children with Down's syndrome, and because so many rapid developments are being made in medicine, genetics, education and the concept of mental retardation itself, we believe much greater efforts need to be made by everyone involved—parents, national organizations and the medical profession—to work together to get good medical treatment, including growth hormone, to as many children as possible, as quickly as possible.

When my husband and I and our friends saw how much our daughter was benefiting from growth hormone treatment, we worried that news of this treatment would follow the old pattern of families being faced with trying to get medical information on their own or learning about treatment from other parents. Already we have seen the anger and guilt of many parents who have tried to get their children tested and failed, or had contradictory test results.

Many have been discouraged from having their children tested because doctors do not know about the current research on growth hormone. They still think short stature is genetic and just to be accepted in children with Down's syndrome. Others feel, as I noted earlier, that short stature just isn't a problem for children who are 'retarded'. Many families simply cannot afford to have their children tested, let alone pay for treatment.

This is largely why we helped form our parents' association in Orlando last year. We believe it is time that we as parents stop blaming the system and start taking some responsibility for learning how the system works and how to get good medical treatment made available for all of our children with Down's syndrome. We would like to suggest that the best thing for parents to do is to work with their local organizations to set up a good network of medical services for their children.

After 25 years of frustration with the medical system, parents of children with Down's syndrome need to realize that the average pediatrician simply cannot keep up with new information the way that a doctor who works with larger numbers of children with special needs can. It is unrealistic to think that every pediatrician can be an expert on every disorder.

Because our children have such a broad array of medical problems, we need to find doctors who specialize in developmental disorders and support them. We need to refer our members to them and ask our pediatricians to refer new babies to them. We must supply them with the abstracts, periodicals, and information they need to learn about what we are doing to help our children. Like many groups, our organization has assembled 'new parent packages' for use in the hospitals, full of information about Down's syndrome and our community resources, including these developmental doctors.

We are talking with our children's hospital about sponsoring a semi-annual

'Down's syndrome clinic', where our children would be seen by all the specialties for screening and testing, in an effort to make sure that no more children would lose their hearing or have uncorrected heart problems. Those who do not have insurance or money for the screenings would be seen free of charge. This is hopefully where many children would be screened by an endocrinologist and referred for growth hormone testing.

We asked pediatric endocrinologists from Orlando and the University in Gainesville to attend this conference, so our parents in Florida would not have to rely on other parents to answer questions about growth hormone research and testing for their children. Instead we will have medical professionals to refer them to so they can hear the pros and cons and the latest research information. With area specialists interested in this topic, perhaps we will even be able to raise money for our own growth hormone research project in Florida so it will be more affordable for our families.

We must ask the developmental doctors and specialists we work with to provide information about new medical treatment and research to area pediatricians at seminars, rather than have parents trying to get this information to each doctor.

Some of these services are already available in a few larger areas of our country. We hope the national parents' organization will work to get a consistent system of medical support available in every area.

Down's syndrome is the leading genetic cause of mental retardation in our country. It is a disorder that causes great harm to our children's bodies and minds. Too many significant new advances have been made in biomedicine and genetics to continue to accept the idea held by the national organizations any longer that it is the only untreatable one. The $75 000 in research grants given by the National Down's Syndrome Society every year is a poor effort to help the more than 250 000 persons in our nation with Down's syndrome. Significant amounts of money need to be raised to support research for our children.

My husband, Tom, and I want to thank Dr Salvador Castells and Dr Krystyna Wisniewski for pursuing growth hormone research for children with Down's syndrome despite the criticism from doctors who base their ideas about Down's syndrome on old information and despite the lack of support or money from either the National Down's Syndrome Congress or the National Down's Syndrome Society. We hope someday these fine doctors will receive the recognition they deserve for the important work they are doing for our children.

We also wish to thank the Institute for Basic Research and its Director, Dr Henryk Wisniewski for supporting their research and this conference. We were told at the National Down's Syndrome Congress Conference in one of the medical panels that the findings at this conference would set the standard for treatment of Down's syndrome with growth hormone, that all of the answers on growth hormone treatment of Down's syndrome will come from this conference. At the very least, we feel certain it will stimulate much more research on growth hormone.

We want to thank all of you who have attended and participated in this conference this weekend. We hope you will throw out the Down's syndrome growth charts and work hard to get this important medical treatment to as many children as possible. There is help for our children. We need more research and more accurate methods of testing.

The children of my daughter's generation will either be the last to suffer from Down's syndrome as we know it, or they will be the first of a new generation who will benefit from medical treatment for one of the major medical symptoms—improper growth. This new generation of children does not have to end up like those who came before them.

I would like to leave you with a statement made to me by the pediatrician on our board, shortly after he began working with our children. He said quite simply, 'These kids are worth saving!'

REFERENCES

1. Cooley WC, Graham JM Jr. Common syndromes and management issues for primary care physicians: Down's syndrome—an update and review for the primary care physicians, *Clin Pediatr* April 1991; 233–234.
2. Kunc N. *Rediscovering Our Right to Belong*, Plenary Presentation to National Down's Syndrome Congress, 20th Annual Convention, 10 September 1992.
3. Wisniewski, KE. Down syndrome (DS) children, postnatal retardation of head and brain growth (birth–five years). Disorders of the developing nervous system: changing view on their origins, diagnoses, and treatment, In *Proceedings of the 1987 Albany Birth Defects Symposium XVIII*, edited by Swann JW, Messer A, New York, 1987.
4. Donnellan AM. *Communicative Functions: Understanding the Learner's Perspective*, Plenary Presentation to National Down's Syndrome Congress, 20th Annual Convention, 20 September 1992.
5. Schmickel R. Recent research for Down's syndrome: summary of address to National Down's Syndrome Interest Group of the AAMD, *Down's Syndrome News*, March 1982.

22 Family Perspectives on Growth Hormone Treatment in Children with Down's Syndrome

CARLOS OBERTI AND JEAN OBERTI
21 California Avenue, Clementon, NJ 08021, USA

We are the parents of an eight-year-old boy with trisomy 21, and were concerned about his growth rate. Our son Rafael was five years old and under the fifth percentile in growth. We were lucky to meet with Dr Wisniewski and later on with Dr Castells and Dr Torrado, for whom we have deep gratitude and warm feelings of friendship.

We have to tell you that most of the trouble we have had in Rafael's life has not been a lack of knowledge of best practices, but rather not being able to obtain those things which we know will be beneficial to our child.

There are three areas that can be improved when referring to the growth hormone treatment: (a) research clearly shows that the benefits of growth hormone treatment in children with Down's syndrome need to be compiled and disseminated; (b) these results need to become part of the regular care of children with Down's syndrome or at least an available option through local pediatricians; (c) this is a costly treatment, there needs to be a way to fund it, so that this does not become another treatment available only to those who have the money to pay for it themselves.

We have spoken to many other parents who have shown interest in the treatment, but when we tell them that we travel to New York every three months to obtain it, their enthusiasm wanes. Endocrinologists, pediatricians and other professionals in our vicinity to whom we have spoken have had no knowledge of this treatment and react as if they do not think it is a very good idea. Both groups ask about the expense and if the results are really worth it.

So, we have a rather large task before us. We do not have a way to document any case but that of our own son. The obvious change for him is an increase in height from below the fifth percentile in height for his age to the fiftieth percentile. This is indisputable, but it is not enough to convince most people that this a life-bettering option.

The obvious strengthening of his muscle tone and increased athletic ability are probably the second most impressive tangible result. The third and probably more difficult to document is the increased learning ability. Lastly, and probably most important to us, is the way our child feels about himself.

Growth Hormone Treatment in Down's Syndrome. Edited by S. Castells and K.E. Wisniewski

If any of you came to this seminar discouraged or lacking the impulse to continue in your research, we hope that has been dispelled. Nothing in life that is worthwhile ever comes easily. You have in your hands the future of hundreds of very special individuals with Down's syndrome.

Parents of special children who care walk very difficult paths with their children, not because of the child but because of widespread ignorance in our society as a whole. Our children never let us down, they give love, and show us the way of kindness and acceptance. But many times even the educational systems fails to give them a chance, and opportunity, the minimal considerations and the respect every human being deserves.

We are extremely thankful and happy to find a group of courageous individuals like yourselves working to improve the quality of life for our children. You are the light that shines and rises above the mediocrity of others that prefer to remain in the Dark Ages.

For us it is very hard to see that most research is done for deadly diseases and adult patients. With the exception of your research, very little is done to improve the quality of life of our children. You bring hope to us. We pray for people like you to continue looking not only at the effects of growth hormone, which we already know helps our child, but also into additional treatments that will enhance memory, promote learning and more importantly prevent degeneration of the brain.

We hope you will look into the future, project yourselves into research, which, in combination with the growth hormone, will take care of the individual as a whole person and make their lives the best that they can be.

23 Summary Discussion

SIEGFRIED PUESCHEL, (moderator)
Salvador Castells and D.A. Snider, (editors)

H. M. Wisniewski: Yesterday, growth hormone (GH) and the blood–brain barrier (BBB) were discussed. The BBB prevents GH from entering the central nervous system (CNS) in mature animals. However, during fetal development in animals, the BBB does not exist. I believe the role of the BBB in GH treatment needs to be re-examined with respect to the fetal and early postnatal period.

V. Sara: There are no data at this time to suggest that GH will cross the BBB at any time. There are no data either that insulin-like growth factor (IGF) will cross the BBB. There is no relationship between serum and CSF levels. Dr Annerén showed a slide yesterday about the level of IGF-2 in children that indicated that there was a fall in those levels up to about three months of postnatal age, which corresponds to the development of the BBB. So prior to three months there is transport from the serum, of course, but after that there is no transport of IGF or human GH (hGH).

S. Pueschel: Whether GH crosses the BBB may not be the most important question. Perhaps the question should be, does GH have to cross the BBB, or the other metabolites, which will stimulate whatever metabolism has to be stimulated in the brain?

V. Sara: The choroid plexus is extremely interesting, especially as it is the source of IGF-2 and IGF binding protein (IGFBP-2) in the brain.

Unidentified: GH does other things beside stimulate somatomedins; there are influences on the metabolism of the body. If metabolic therapies have some effect on these children, the effect of GH may be through drops in the blood urea nitrogen or changes in amino acids and fatty acids; these things are transported. The delivery of metabolites is altered by GH. There are many other metabolic events happening during GH therapy and they need to be examined.

Z. Laron: There is some clear evidence of immediate effect of IGF-1 on somatostatin in the brain. When we give one intravenous bolus of somatostatin, we get a fast reduction in thyrotropin. If this happens through central somatostatin, then we have to imply that the message gets through to the brain somatostatin. At the present time, we don't have evidence as to how the interaction comes about.

D. Sandberg: I am a psychologist and focus heavily on the effects of short

Growth Hormone Treatment in Down's Syndrome. Edited by S. Castells and K.E. Wisniewski

stature in my research. I don't think we can conclude from the clinical literature on short stature across many conditions that it is short stature that causes poor adjustment. Second, the evidence indicating that GH reverses this liability is also not very compelling and it is controversial. There is some indication that there can be significant disenchantment or disappointment after the patients have undergone treatment. But this is based on very poor data, and most of it predates the introduction of biosynthetic GH. We find now that the populations of patients undergoing synthetic GH treatment are quite different from the ones that were studied in 1985 in the USA.

R. Levy: I think you may have to move away from not wanting to deny Down's syndrome children basic care or for that matter GH if they have GH deficiency. We should recognize the fact that about 30% of the children who are now in the National Collaborative Growth Hormone Study and are being treated for short stature are not characterized as having GH deficiency. It has been brought out previously in this symposium that Turner's syndrome is more of a GH resistance than it is a GH deficiency and there are relatively few patients who test that way. It comes down to the question of whether Down's syndrome children should be treated like any other child, and if they are short, does that mean they deserve to have GH as much as a person with a dysmorphic syndrome or something else?

G. Annerén: Parents here have asked me if I am against GH treatment in Down's syndrome? And I say no. But I do say that we do not know the final height and we know nothing about brain function. If we are talking about a child with Down's syndrome, three standard deviations below the mean on growth, I would say treat her or him like a normal child, but if you are talking about treating all children with Down's syndrome, I say no. We need to study this more and be patient.

S. Pueschel: We need more good research, and I emphasize the word good. We need to know where these treatments lead us, what is the longitudinal expectation? What does it do to brain growth and intellectual function, and what are the side effects?

P. Wilton: I disagree with those who say we should plot the height of Down's syndrome children on the growth charts of normal children. It is important that we use a disorder-specific growth chart. I agree we could treat Down's syndrome children already today with GH if they prove to be GH deficient.

T. Sharav: I think that one has to be very careful in assessing if it is GH that helps any of the variables we assess in children. First, we don't know if the Down's syndrome group is affected by the secular trend, as is the normal population. Do familial factors, such as tall parents of Down's syndrome children, have tall children? We know infections affect the size of children. Many children with Down's syndrome have feeding problems. I think the same may be true of intellect development. Down's syndrome is a heterogeneous population. We are getting a broad distribution of intelligence within the Down's syndrome population, though lower than the normal population. To isolate one factor, GH, and say because they got hGH and then they grew and got smarter without taking into consideration other factors is unwise. Also, we must start treatment in early infancy. We need more longitudinal data. Most of what we are seeing is cross-sectional.

Unidentified: I believe that the secondary conditions that Dr. Sharav mentioned are the very things the parents are trying to alleviate. We have had stimulation programs and health programs, and our children still are not growing.

G. Merriam: I would want to agree with the point of having good clinical trials of GH, and I also want to underline the word good. A very strong argument was made earlier by concerned parents about the value of general medical care and social stimulation in improving the situation of these children. In this situation, the parents who seek these treatments out are often the most motivated, the most diligent, and the children may do better independent of whether they receive GH. If we are going to undertake such trials, they must be structured with proper controls, so that we know what effects are due to the parents or medical system intervention and which ones are due to GH itself. The strongest form of such control is randomization, otherwise people are going to be putting in a lot of effort and a lot of expense and we still may not know the answer.

Index

Note: page numbers in *italics* refer to figures and tables. DS refers to Down's syndrome throughout.

Index compiled by Jill C. Halliday